FORCES OF LIFE

THE BOTANIC MAN

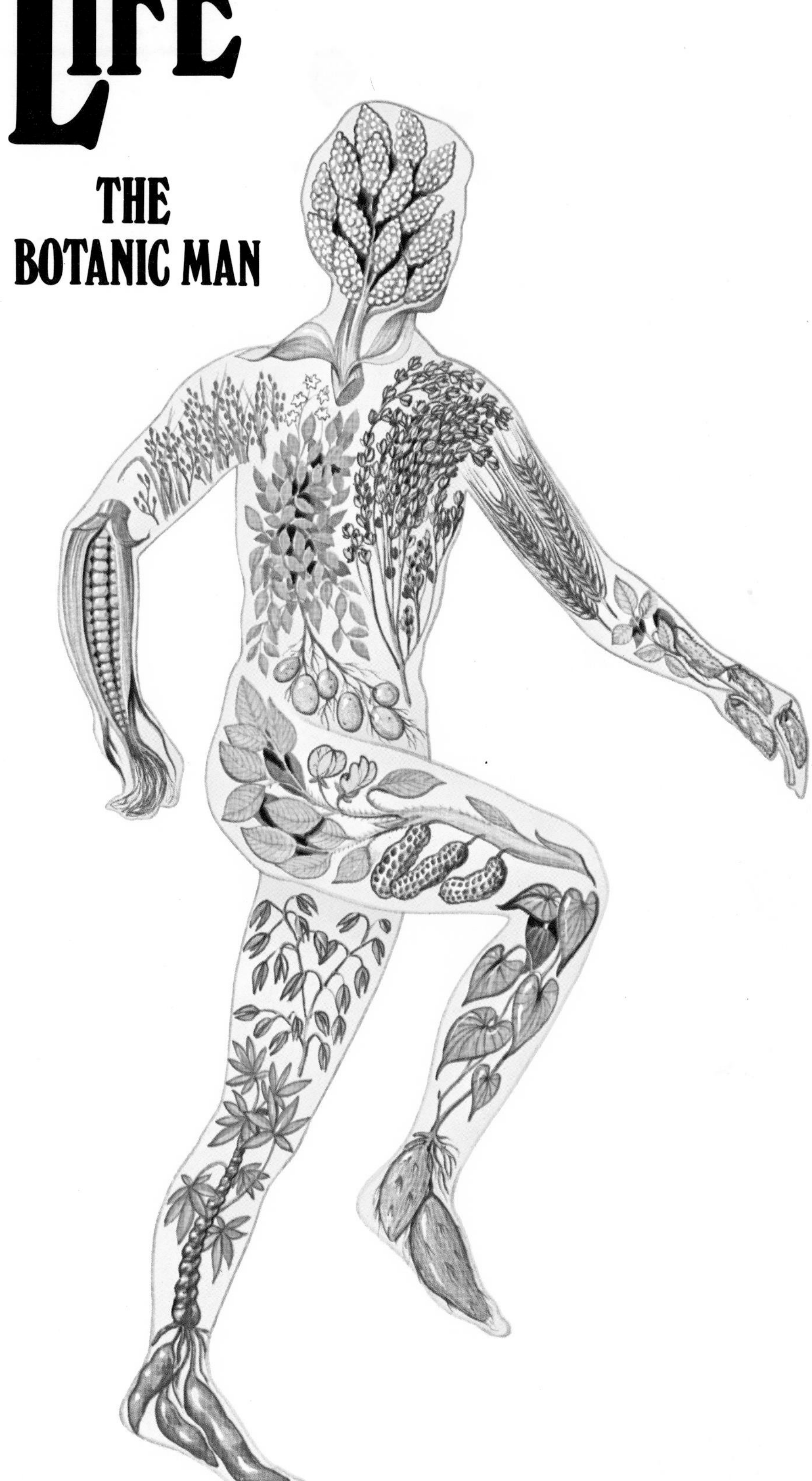

FORCES OF LIFE

DAVID BELLAMY

THE BOTANIC MAN

Crown Publishers Inc., New York

DEDICATION

During the making of the first of the series of television programs which goes with this book I was privileged to stand close to humanity. For a few precious hours I ran barefoot through the most affluent natural society on earth and came to know something of the joys of a child of the forest and a way of life which is sadly almost extinct.

The child is Obé, daughter of Apa and Camemono, members of one of the last groups of Wodani, a meek people who still live as part of the forest of the Upper Amazon.

It is in the hope of some meaningful future for mankind that I dedicate this book to Obé and her people.

First published in U.S.A. by
Crown Publishers, Inc., 1979

Library of Congress Cataloging in Publication Data

Bellamy, David J
The forces of life.

1. Evolution. I. Title.
QH366.B44 575 78-16360
ISBN 0-517-53529-7

Contents

Artwork by The Tudor Art Agency Ltd.

INTRODUCTION

Botanic Man is the culmination of a lifelong interest in people, especially children, some twenty-five years of academic research all over the world and teaching at the University of Durham, England, and ten years' working through the media of books and television to introduce people to the fascination of the living world of plants.

This book concerns you for you are Botanic Man—it is about your past, your present, and your future.

It tells how the dead world became covered with a mantle of living green, upon which the productivity of every member of the animal kingdom ultimately depends, both for the food he eats and the oxygen he respires. How man's meteoric rise to world domination has itself been entirely dependent on plants: first through the agricultural revolution that gave us the power to control, at least in part, our food supply; then through the industrial/medical revolution, energized by fossil fuels, which were the excess of past ages of photosynthesis when green plants stored more energy than their dependent systems could make use of. How the future success of man must depend more and more on the sun as a source of energy and the plant kingdom as a source of raw materials.

In essence, the book puts man's past, present, and future into the perspective of evolution as seen through the medium of the plant kingdom and comes to a positive conclusion.

Plants evolve within structured communities, which make maximum long-term use of the potential offered through a program of ordered recycling of all nonrenewable resources. In such communities there are no top dogs; the fittest are those able to play a particular role within the life of the community, and only those survive. As such, evolution becomes a much more positive directive for man in the latter part of the twentieth century than the rat-race philosophy so often popularized through the tooth-and-claw world of the animals.

"Fit to play a role in the community of life," this is the true philosophy of evolution, a philosophy that can earn for man the final stamp of evolutionary fitness—survival.

The world is a very wonderful place in the caring hands of Botanic Man—and it will continue to be.

I am both a botanist and an optimist—that's why I wrote this book.

DAVID BELLAMY

Chapter 1

Green print for life

All flesh is grass, yet today many of us live with our roots so far removed from agriculture that this oft-quoted statement has little real meaning.

This chapter challenges you to live with nothing but your own wits in the most productive and diverse natural community in the world – the rain forest which clothes the banks of the River Amazon.

It gives you the low-down on the more important aspects of the other members of the community of nature with which you would have to learn to live and it warns of the strict limitations you would face – limitations which would determine your ultimate success as it determines that of the local indians who evolved to live as part of the forest.

X, you are going to be put to the ultimate test of survival. Take your family with nothing except the skins you stand up in and live off the land! How do you think you would get on?

Without prior knowledge and planning, the answer would certainly be not very well. The first problems would be where and when; where your new 'home' was located and at what time of year the test was to begin. If you could choose the spring or early summer you might have a fighting chance, but if it was autumn or winter you would be in for a lot more trouble. Protection, water and food would be your immediate requirements. So, let us choose a spot on the equator where there are no seasons, where it rains at regular intervals and where there is abundant plant and animal life. Here is some information about the habitat.

The diagram (right) gives the vital statistics of part of the Amazon basin, average temperature 26·2 degrees Centigrade (79·2 degrees Fahrenheit); there should be no problems with goose pimples and plenty of rain to water the plants. These particular climate diagrams were dreamed up by a scientist called Heinrich Walter who wanted to be able to compare the climate of those parts of the world which have similar types of vegetation. To the trained eye, that diagram says tropical rain forest and even to the untrained eye it is pretty impressive stuff, and that is going to be your new bit of real estate, your life support system.

One of the first descriptions of tropical rain forest was written by no less a traveller than Christopher Columbus, who visited Española (an island in the West Indies, now known as Hispaniola) in 1493 and enthused about its luxuriant forests. The first real scientific description was not made until over 400 years later when a German botanist, Schimper, published the classic description of *Regenwald*, which means rain forest.

'Evergreen, hygrophilous, at least 30 metres high and often much higher, rich in thick-stemmed lianas and woody as well as herbaceous epiphytes.'

Between the time of Columbus and Schimper, the Amazon basin was discovered and many explorers and naturalists made their way into its

5·2 million square kilometres (2 million square miles) of forest. Many lost their lives, some made their fortunes, but all added to our knowledge of the most productive wilderness on land.

The first impressions of the great forest were, and still are, of an enormous wealth of life with giant trees reaching up into the skies, blotting out so much of the sunlight that few or no green plants can grow on the forest floor. It is nothing like the storytale jungle of films and television in which the explorer has to hack every step of the way through dense undergrowth, putting to flight flocks of shrieking parrots and keeping a wary eye out for venomous snakes which lurk behind every fallen log. A walk through virgin forest is more akin to a stroll through an English beechwood or better still down the aisles of some vast cathedral. You are aware of the cool brown shade, a smell of damp decay, and all around you the pillar-like tree trunks disappear into the arched void above. The trees are not crowded and there is ample room to walk between, and the majority of the trunks are very slender, being less than 1 metre (3·3 feet) in girth. However, here and there enormous trunks sprout from the ground; a 10-metre (33-foot) girth is not rare and there are authentic records of incredible trees, the trunks of which are more than 17 metres (56 feet) in circumference. It is almost as if they are part of another world–a green heaven held high above the main canopy.

The semblance of a cathedral is enhanced by the hallowed silence that reigns supreme for much of the day. If you stand quite still and listen you will find that the silence is, in fact, a background hum of millions of busy insects and it is possible to latch on to the ordered beats of one that is slightly out of time with the rest. When this busy silence is broken by some harsh cry from the dim distance it comes as an immense shock and on many occasions I have stood, hairs on the back of my neck erect with anticipation, waiting for the end to come. The trouble is that you never really know what it was that broke the silence and guessing can only make it worse.

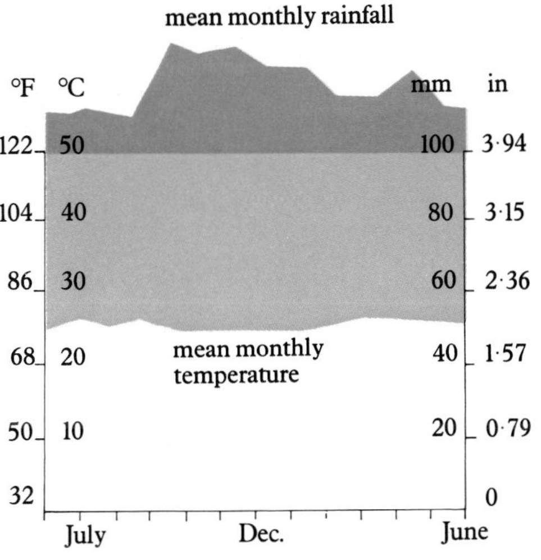

The diagram summarizes the climatic data collected at San Carlos de Rio Negro which is situated 110 metres (360 feet) above sea level over a period of 8 years. The mean annual temperature is 26·2°C (79·2°F) and the mean annual rainfall is 3521 millimetres (139 inches). The lower line records the annual march of the monthly mean temperatures and the upper line similar data for the mean monthly rainfall. The former hardly varies throughout the year and the latter is so high throughout the year that the scale had to be reduced by one-tenth on the top green portion.

The noise of the forest comes to a climax just after rain and as the sun goes down, as if these changes of environment awake a sleeping life within. A walk at night is in some ways much more exciting because then many of the larger animals are awake. There is warm inky blackness all around and the noise of some large animal moving through the forest. What can it be? A giant Boa Constrictor, angry wild pig or panther? A flick of the switch and the torch shatters the illusion of fear as an armadillo shakes its quills and rattles away, intent on its own meal.

The lack of life, both plant and animal, down at ground level is mainly due to the structure of the forest itself, which is a direct response to the environment in which it evolved. Survival depends on getting your fair share of the energy of the sun. Here, where it is warm enough and wet enough all year round to support the growth of plants, it is 'all systems grow' upwards and anything that gets left behind in the race to the light is in for real trouble.

When naturalists like Henry Walter Bates and Richard Spruce first walked through the forests of the Amazon, they were amazed to find gigantic trees bearing flowers like daisies and violets which back home in Britain were only found on delicate herbs. They soon realized that in these forests there was no place for ground-living herbaceous plants. For a plant family to thrive in the wet humid tropics it had to get up in the world of the trees. It was a German botanist, Gams, who formalized our ideas about the importance of light in the rain forest when he classified its myriad plants into functional groups called synusiae, plants which play similar roles in the forest.

Plants, unlike animals, usually have their living cells enclosed in a tough box made of cellulose or some other structural sugar. Green photosynthetic plants contain the pigment, chlorophyll, which is the key substance in the process of photosynthesis, fixing light energy and storing it in the form of sugars.

The mechanically independent plants (trees, shrubs and herbs) are those which can support themselves in the thin air.

Trees are large plants, the stems of which grow thicker as they grow taller by producing tough supporting tissue, called wood. Wood can come in a variety of different forms–ultra-light, easy-to-cut balsa wood at one end of the density scale and dark, heavy ebony, which is almost impossible to chop, at the other.

Herbs, by definition, never produce any massive wood, which severely limits their size. It is, however, a fact that even the mightiest tree must start its life on the forest floor as a herb and thereby hangs one of the most important stories of forest life. There are, of course, some freaks among the herbs, and perhaps the banana is worth a mention. These large and important plants produce no woody material. Their support lies in the strength of their massive leaf bases which encircle the stems; in fact, it could be said that the leaves hold themselves up towards the light.

Shrubs are usually woody perennials and you can really put your own size limits on them. To complicate matters it is worth remembering that if all the trees are herbs when young, they probably pass their adolescence as a shrub.

Once you have sorted that lot out you have begun to understand the self-supporters. Fortunately, the tropical rain forest got itself sorted out many millions of years ago and,

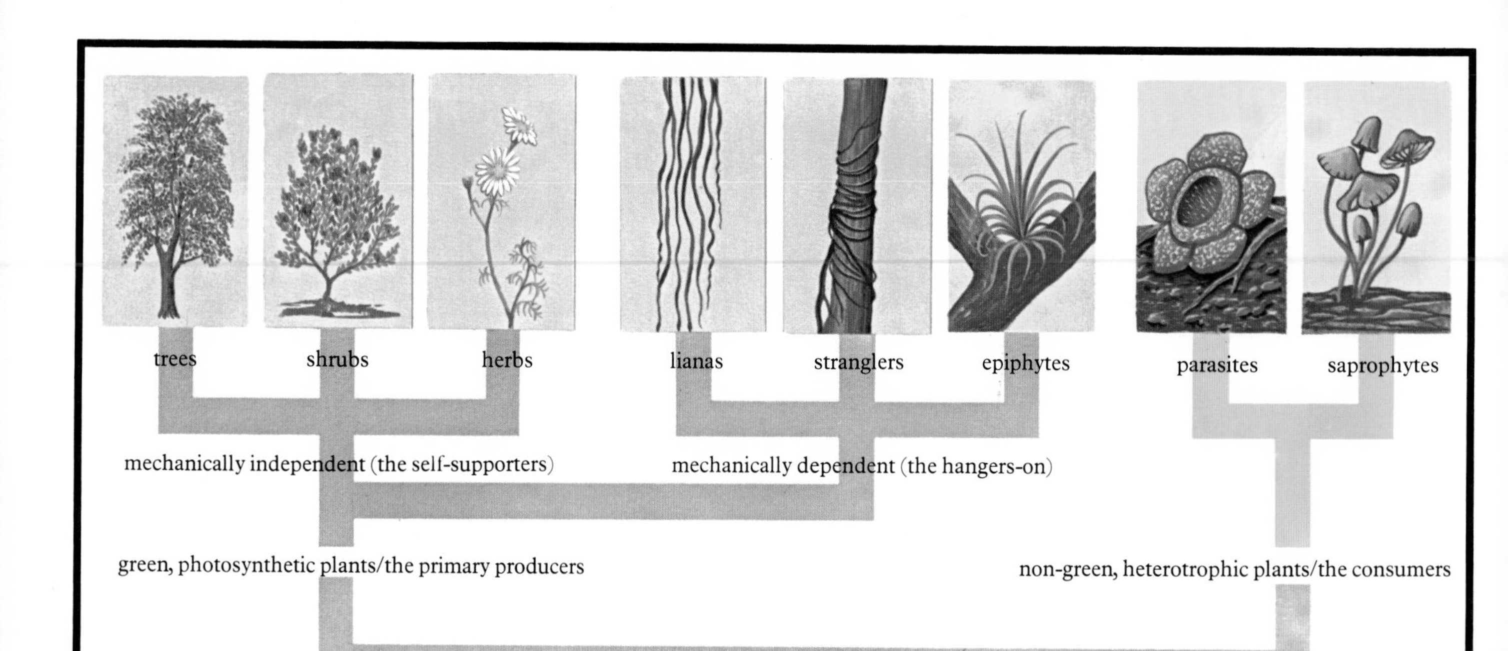

Synusial classification

The forests of the Amazon basin rank among the most diverse types of vegetation in the world. We do not even pretend to know how many different species of plant grow there and no botanist has ever lived who could put a name to more than a few per cent of them. That is one reason why Gams dreamed up his simple classification system, grouping them into functional units. The question of the great diversity is discussed again in Chapter 6, but whatever the explanation, at present every one is a winner.

How a young herbaceous stem starts to turn into a woody trunk

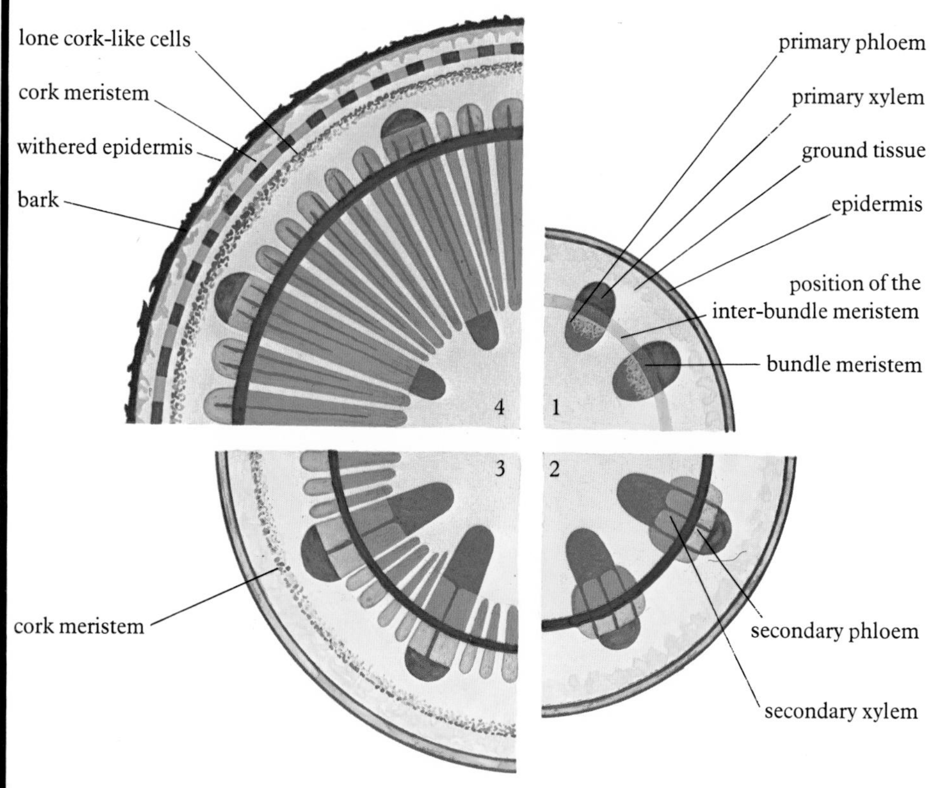

Sector 1 The young stem consists of an outer protective coat (the epidermis) which encloses a living ground tissue through which vascular bundles conduct water (via the red xylem) and sugar and other manufactured substances (via the green phloem) around the plant. As the plant grows taller the stem starts to get top heavy and extra strengthening material is required.

Sector 2 A new (secondary) meristem, the inter-bundle cambium, appears and starts to divide producing secondary phloem and secondary xylem (wood), enough to support the growing plant and keep its parts well supplied with everything they need.

Sector 3 As growth continues the outer layers come under more and more strain until eventually the protective epidermis must rupture. Before this happens another secondary meristem develops within the stem.

Sector 4 This is the cork cambium or phellogen and it divides to produce a new complex protective coat called cork. The old epidermis is now dead and slowly decays away. As the secondary wood develops, the transverse medullary rays elongate. Their function is to transport material from the side to the middle and vice versa.

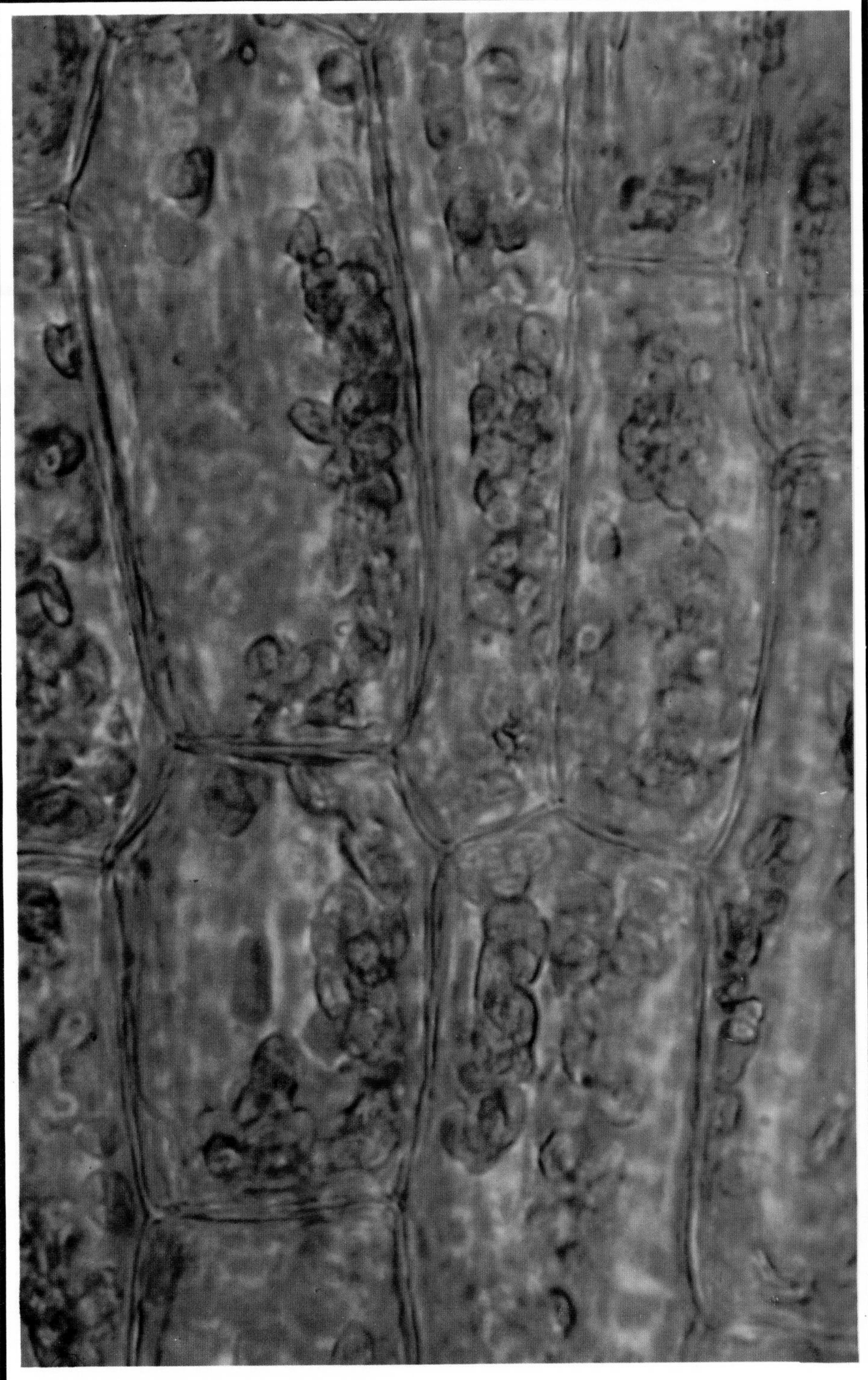

Plant cells were first seen by Robert Hooke (1635–1703) when he looked at a piece of cork under a microscope. He described them as 'many little boxes' and marvelled at the fact that there would be 'above 1200 millions in a cubic inch'. The living cells in the photograph are from the leaf of a water plant magnified 1000 times and show the living cell wall, the green discoid chloroplasts and large vacuoles surrounded by living cytoplasm.

In transverse section it is possible to see the differences in the structure of the wood of balsa (*top*) and ebony (*middle*), differences that are reflected in their properties and hence their uses.
Bottom. A section of part of the stem of a liana (× 15). The black tissue is the wood which gives the stem its strength and ability to transport water. The white wedge protruding down into the wood looks and acts like a hinge, allowing the stem to twist.

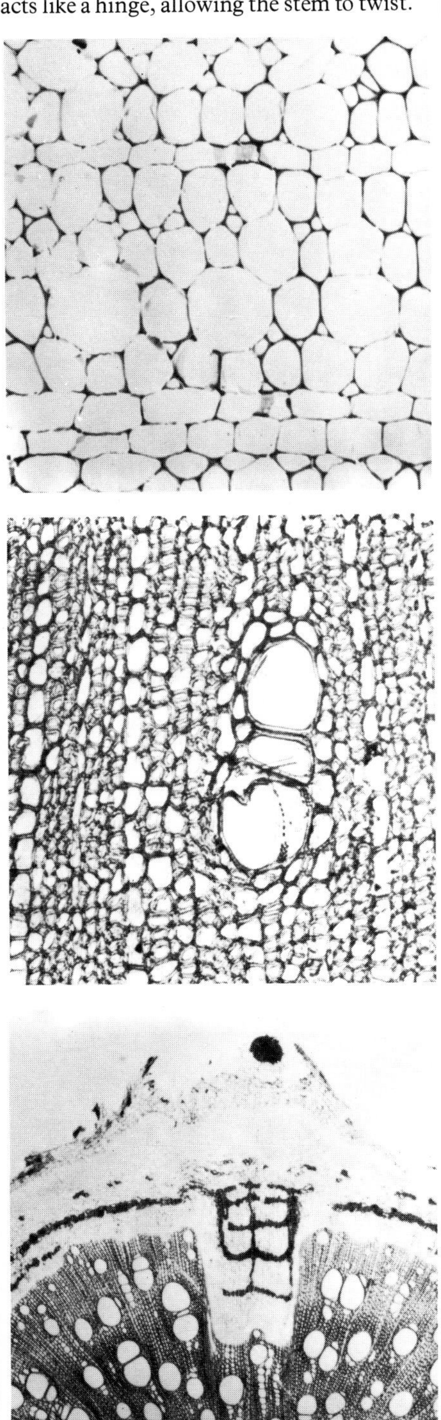

Above What a herb, and what a spike of flowers! The apparent erect stem is formed of the closely overlapping sheathing bases of its large leaves. The banana is a native of India and the East Indies, and was taken to Africa in around 1500. Today it grows both cultivated and wild throughout the tropics and forms an important part of many peoples' diet.
Right The interior of virgin rain forest as seen from the River Napo, a tributary of the Amazon. Although such views may well be atypical of the main areas of forest, which receive much less sunlight, they are the best the average botanist ever gets. At least it allows us to see some of the other members of the forest flora apart from the trees, such as the pendant lianas.

although it comes in a whole variety of forms, the most important features of its structure are summarized in the diagram (p 8). Each stratum of self supporters is holding its leaves up to intercept the energy of the sun.

In 1 square metre of tropical forest there is 35 kilograms (77 pounds) of wood holding up 5 kilograms (11 pounds) of leaf which together fix an incredible 84 million joules (62 million foot-pounds) of energy each year, enough to keep the largest family going. Apart from supporting their own chlorophyll they also support the mechanically dependent members of the forest, which, according to Schimper's description, are one of the main trademarks of the rain forest.

Lianas did not evolve simply for the swift transport of budding Tarzans, although I must confess that I have never had much success with this form of forest progression. Lianas are really trees that have lost the power of self support. Starting their lives off as herbs, in the absence of support trees, they can grow into quite ordinary looking shrubs which flower, set seed and fruit producing a new generation of would-be swingers which rapidly cover any clearing in the forest. However, as new real trees gain a foothold, the lianas hitch a ride and grow up with them, scrambling from one support to another by twisting and often hanging on by means of hooks and spines. In this way they keep their roots firmly anchored in the soil and their leaves up with the best at the top of the canopy. The secret of their success is neither in the presence nor the absence of wood, but in the way in which it is laid down in the stem. As the stem grows, alternating bands of wood are laid down, each one separated by wedges of softer packing material. In this way the plant body has both strength and flexibility so that it can twist and turn, keeping up with its woodier co-dwellers. There are lianas on record which are in excess of 300 metres (980 feet) in length, at least three times as tall as the record rain forest tree.

It is a fantastic feat of plant engineering to have produced a system with sufficient strength and malleability to grow to that length, and to keep the leaves at the top end supplied with water from the roots, and the roots supplied with energy-rich sugars from the leaves. The water transport of one liana is developed to such an extent that thirsty travellers can cut a length of the stem and slake their thirst with sparkling pure water.

If the lianas take the prize for the longest plant, then the stranglers

must rank among the weirdest. Living a life style not unlike their sinuous cousins, the stranglers send down great masses of new roots to tap the resources of the soil. In time, the strangler completely encloses the support tree in a living coffin inside which the tree may die and rot away.

It is sometimes difficult to differentiate between stranglers and lianas, for the latter can grow to such an immense size that they overstrain the strength of the support tree and bring it crashing to the ground. A newly fallen tree in the forest offers a rare opportunity for less agile botanists to take a close look at the most abundant of all the hangers on, the epiphytes. From down below, a pair of good binoculars and a crick in the neck reward you with a tantalizingly distant view of weird green shapes and exotic blooms.

In a favourable situation, a tree can support almost twice its own crown weight in epiphytes, together with an equal amount of aerial humus perched precariously in the crotches of all its branches. Large epiphyte nurseries are another cause of the downfall of the forest giants– a gust of wind in the wrong direction can bring a top-heavy tree rapidly to rest. Perhaps the most dangerous time for forest life is when the wind starts to blow.

To be a successful epiphyte a plant requires a number of attributes and among the most important are light seeds which can be lifted up by the more gentle updraughts of the forest air. Once in position the next problem is to get fixed on to the support and there to obtain sufficient water and mineral salts for healthy growth. The first comers must face the greatest problems in this respect, since they will only be able to gain a safe foothold where rainwater flows down the trunk, bringing with it nutrients washed down from the canopy. As the community of epiphytes builds up it becomes easier because the whole thing is a cumulative process; the humus produced by the plants already in position helping to provide a damp compost on which other seeds can germinate.

Nevertheless, it is easy to see that the main problem faced by the epiphytes is one of water supply because many of them have attributes of cacti and succulents, plants which we usually find growing in the drier, not in the wettest, environments on earth. Just to rub the point home, an abundant member of the epiphyte flora is the Christmas Cactus (*Zygocactus truncatus*), so well loved as an indoor plant and so much abused when family holidays plunge it into a prolonged period of drought. Living high up beneath the thick canopy of leaves, water supply to the epiphyte nurseries can be just as intermittent. Therefore, the majority of them have thick fleshy leaves or even leaf-like stems, which have the

Left The dreaded strangler strikes again.

Below Two epiphytic bromeliads in full flower. A canopy tree has toppled over and has brought the epiphytes down to photography level. The overlapping bowl of leaves which forms the plant's own water reservoir and the large and flamboyant spike of flowers can be seen to advantage.

ability to shut up shop when water is in short supply and stolidly hold on to all the water they have got. As soon as more water comes their way they can quickly get underway and make up for lost photosynthetic time.

The roots of most epiphytes also show some fantastic adaptations. Like all roots, their delicate growing points are covered by a tough cap, the usual function of which is to protect them against abrasion in the soil. Dangling down in the dry air the cap takes on a new role, slowing down the process of desiccation. The young roots are often green in colour and this adds to the plant's photosynthetic surface, but not for long because as the root ages it soon becomes covered with a tough waterproof coat. In some cases this whole outer layer or layers of the aerial root change into a special tissue called the velamen which seals the living root from the dry air outside preventing water loss. So effective is the velamen that the plant would not be able to take up water if it were not for the fact that wherever the roots are in contact with the support plant no velamen develops and it is through this region that the epiphyte obtains the bulk of the water it requires.

It is not only the epiphytes that display unusual root systems. In fact, some of the largest trees produce the most bizarre root structures of all. Normally shoots grow upwards towards the light and away from the pull of gravity, while roots grow down into the damp darkness of the soil. If the epiphytes suffer from too little water the trees can suffer from too much down in the soil, and if the soil is 'full' of water then there is no room for oxygen-rich air and the roots die. For this reason many of the trees have only very shallow rooting systems which makes for instability, and the larger you are the easier and harder it is to fall. It is for this reason that the roots of many trees grow out at an angle to the force of gravity to produce immense buttresses that appear to prop the trunk up. Real prop roots may grow from the branches of the mature tree. Growing down they produce a structure not unlike the frame of a wigwam, again adding support just where it is required.

Roots growing out at an angle may not seem too bad, but there are others which completely disobey the rule of the root when it comes to gravity. In the swampiest parts of the forest, especially along rivers and estuaries, the roots of certain plants seem to go haywire. Some grow up above the soil and then bend back down again, so much so that they look like rows of knobbly knees sticking out of the ground. In the most extreme cases the branches of each root grow upwards and look like rows of fingers sticking up into the dry air. A close look at both the knees and the fingers will show that they are both well supplied with breathing pores (lenticels) which do just that, allowing these roots to breathe and stay alive.

So it is that each of the green plants is adapted by its own way of life to fill a particular role within the forest. Together they, the primary producers, make this type of forest the most productive found on land.

The non-green heterotrophic plants live off this affluence, off the crumbs that fall from the table of the green plants. These are plants which can grow in the darkness of the forest floor for they do not require light in order to make their living.

Parasites live on other living plants and they range from the bacteria and fungi which cause plant diseases (yes, plants do suffer from such things) through to the gigantic *Rafflesia arnoldii*, the plant with the largest flower in the world. Although this giant is only found in the rain forests of south-east Asia, other members of the family are widespread.

The saprophytes living on dead and decaying material are just as diverse, and one of the delights of a forest walk are the myriad toadstools and other fungi, which are the main agents of decay. Without the green plants there would be no forest, and yet it is just as true to say that without the decomposers the life of the forest would soon come to an end.

In this modern world of ours it is not an unusual thing to hear that certain municipal workers have gone on strike, leaving a city festering in its own refuse. The strike must be ended for the health and welfare of the community at large. The work of the vegetable trash-can brigade is just as vital to the welfare of the forest. It is not just a simple matter

Overleaf The most extreme form of root behaviour is seen in the breathing roots of the mangrove *Avicennia*.

Meristems

At the tip of every shoot and root there is an area of cells, all of which are capable of all the core life functions, including division. These are the primary or apical meristems, the main function of which is to produce new cells. These will in time undergo chemical changes which will fit them to play various roles within the primary plant body, allowing the plant to grow in length only. The shoot apex buds off leaves which help to protect the delicate growing tip while the root apex produces a special protective root cap. The magic of the meristems lies in the fact that at each division all the contained chemical information is duplicated so that each daughter cell has all the potentialities of its parent.

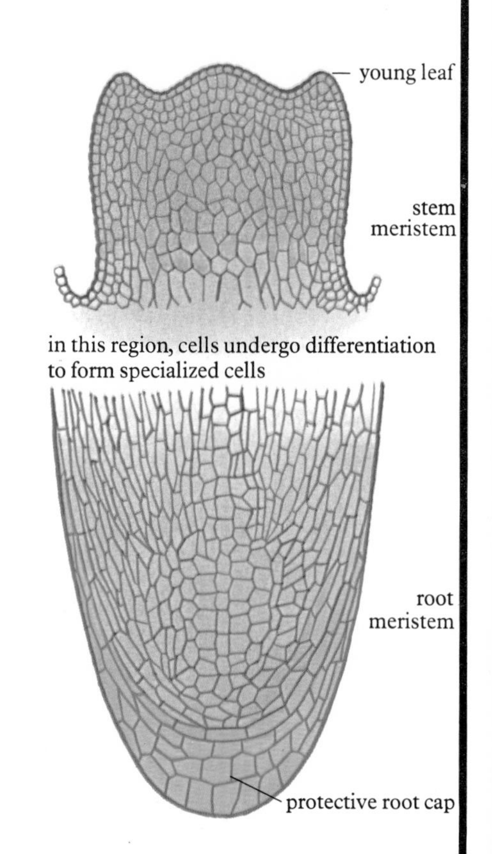

of clearing up the leaf litter. It is much more important than that because the constituents of the leaf litter (which may be as much as 3·5 kilograms per square metre (0·7 pounds per square foot) per year) must be recycled if the forest is to continue in production.

Large as the forest may look and productive as it may be, certain raw materials are in desperately short supply. The minerals needed for healthy plant growth come ultimately from the bedrock from which the soil is formed. In the stable conditions of the humid tropics the process of soil formation has been going on for many millions of years and today the mineral-rich bedrock is way down beyond the reach of the roots. The main bulk of all the useful minerals like potassium, phosphorus and nitrogen are locked up in the standing crop of vegetation. It is for this reason that new growth must wait until the process of recycling makes the supplies available for re-use.

There is, however, the added problem in such a rainy climate that any minerals released on the death of a plant could be easily washed away out of 'reach' of the living plant roots. This is where the decomposers play another important role in the life of the forest. Each leaf which falls to the forest floor represents a potential energy source for a whole host of decomposers and they move in, rapidly breaking down the organic material. The minerals are not released; they pass to the fungi which ramify through the soil carrying them directly to the surface of the plant roots, a recycling plant from which little or nothing is lost.

In the same way, the branches and the trunks of the trees are dealt with but here the first step in the chain of decomposers is an animal, called a white ant or termite. Wood is very tough, being made of cellulose, and cellulose is made of glucose. All athletes know that pure glucose is a good source of instant energy and a glucose tablet just melts (dissolves) in the mouth. It, therefore, seems a strange choice as a structural material. However, once glucose has been polymerized (that is, a lot of molecules have been stuck together) to form cellulose it is almost inert, which means that it is not susceptible

The fungi

Why is a fungus like an iceberg? Because you can only see a small part of it sticking up into the dry air.

The part of the fungus we see is more often than not the fruiting body, the bulk of it being out of sight below ground. The underground portion consists of delicate tubular cells called hyphae which grow through the substrate seeking out supplies of organic matter on which they feed. At certain times of the year the fruiting bodies are produced and it is then that the expert mycologist (someone who studies fungi) can begin to identify them.

Fungi come in three main forms. The Phycomycetes include all the moulds which grow on organic waste. These are the smallest fungi and many of them are parasites.

The Ascomycetes are my favourites because they include yeast which not only puts the alcohol in the bottles and the chemical industries of the world but also the rise in our daily bread. The trade mark of this group is a spore gun called an ascus which fires eight ascospores under pressure.

The Basidiomycetes include the majority of the fungi which make an autumn walk in the woods such a fascinating experience and add a number of unique tastes to international cuisine. Their spores (basidiospores) come in batteries of two or four, each fired from its own launch pad (basidium); the latter line the gills or the pores of the fruiting body (the bit we see).

Basidiomycetes (e.g. mushrooms and toadstools)

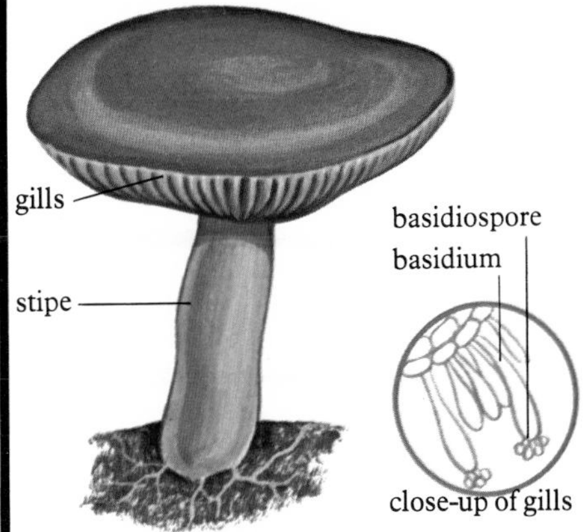

Ascomycetes (e.g. cup fungi)

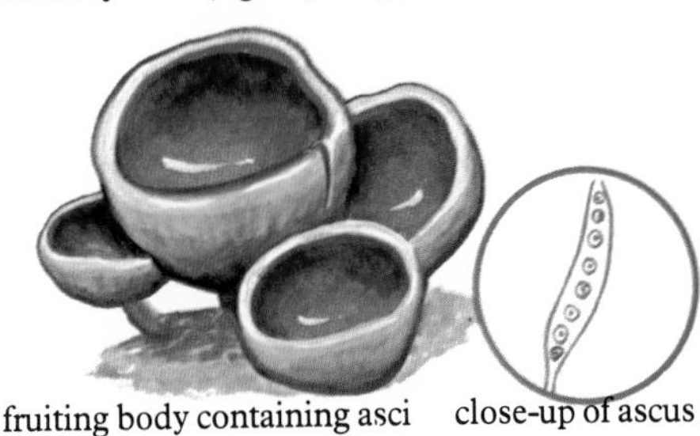

Phycomycetes (e.g. moulds)

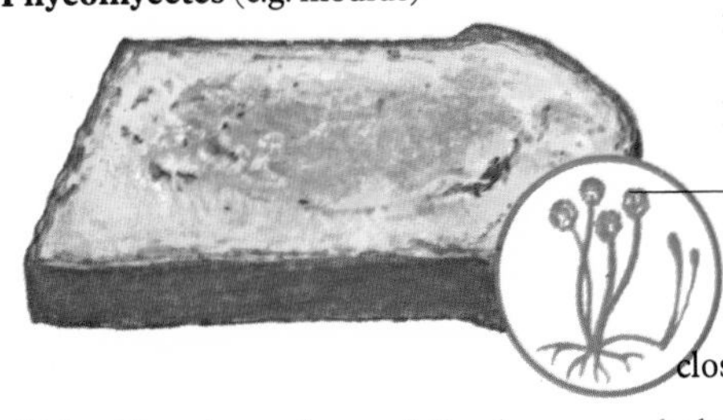

Below Fungi growing on fallen leaves mark the end of one cycle of life and the start of another.

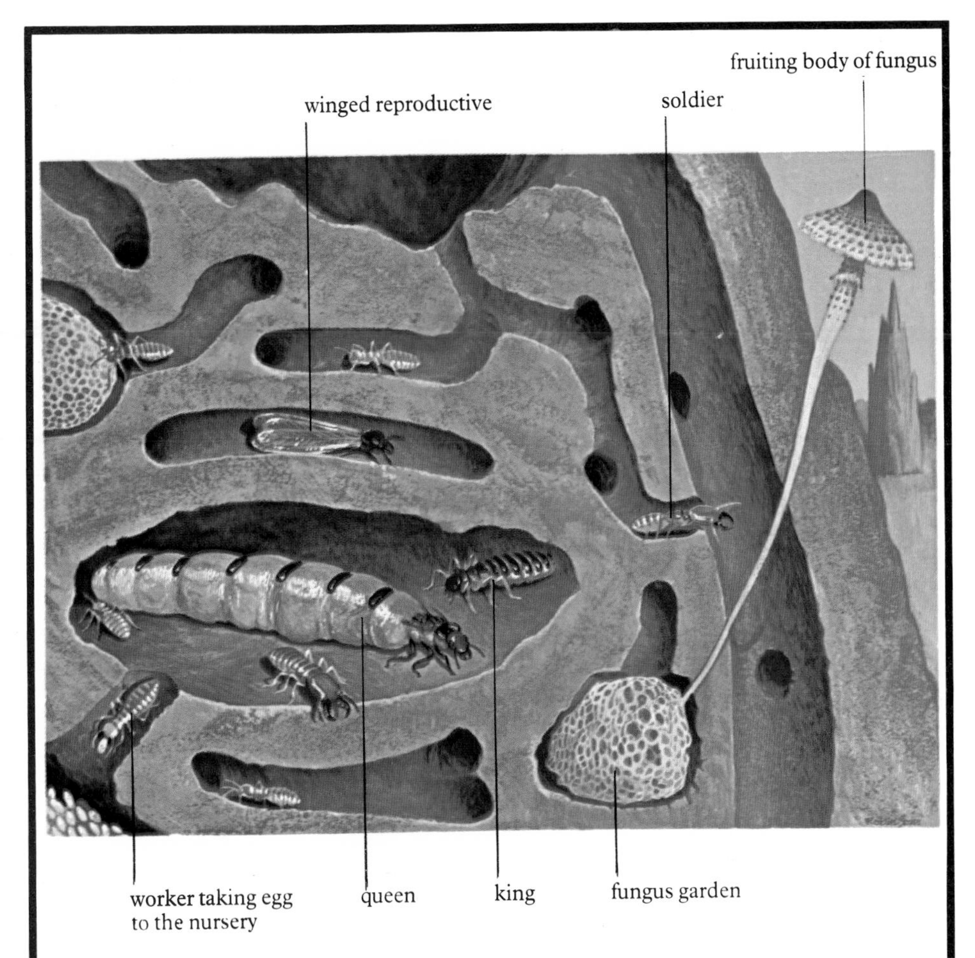

The termites

Throughout the tropics the termites play a key role in the army of decomposers which keep the vegetation clear of natural litter and waste and put the all-important minerals back into cycle.

Termites are social insects and in their society everyone has a job to do. The picture shows an African species which builds gigantic nests (termitaria) of rock-hard soil. No matter where they live or how they build their nests, when it comes to social organization they are the tops. The life of each colony revolves around the queen whose job is to carry on laying eggs, one for every second of her regal life. The eggs develop into either workers or soldiers whose titles accurately describe their roles in their particular social whirl. The workers build the nest and keep it clean and in repair, going out only to forage for the woody refuse that feeds the colony. The soldiers guard the nest and the workers while they are out on a foray. Nothing goes to waste for even the droppings are carried to special galleries within the nest where they are used as compost heaps on which special fungi are grown, thus supplementing the food supply. Under certain conditions some eggs give rise to a third type of individual, a winged reproductive, which leaves the nest to mate and form a new colony. I have a feeling that our own society has a lot to learn from the termites.

Left Buttress roots, a termite colony's nest on a young tree, and an even younger tree waiting to make its bid up towards the light, are all part of the web of life of the tropical rain forest.

to chemical attack except under very special circumstances. Also, the cellulose in wood is impregnated with lignin which helps to make it waterproof and rot proof, one reason why wood is such a good building material. This is, however, not so true in the tropics because termites can digest wood, breaking down the cellulose, releasing the sugar and thus obtaining the energy they require.

If it were not for the termites the fallen trees would take a very long time to decay away. During this time the nutrients would be useless and as the minerals were slowly released they could be leached away by the rain before being re-used. The termites aid rapid decay, opening the way for the other decomposers. It is not an unusual thing to see the 'ghost' of a fallen trunk picked out by a row of young saplings living it up on the ready supply of fertilizer.

It is not only in the cycle of decay but throughout their life cycle that animals come to the aid of the plants, and never more spectacularly than at the time of pollination. Many of the forest flowers are adapted to attract and hence be pollinated by a range of animals from the smallest insects to the largest of the humming birds, and even by bats. The attraction mechanisms may be nectar, odour, colour and shape of the flowers. The range of mimicry between the flower on the one hand and the animal on the other is a constant source of wonderment both to the layman and the scientist. This is, at the moment, a growth area in the biological sciences and, as more workers are attracted to this exciting field, new and fascinating animal/plant interactions are coming to light. Among the most peculiar are the myrmecophytes, which means ant plants.

Not all the insects of the forest are helpful as far as the plants are concerned and many forest trees are just as susceptible to a bad attack of caterpillars as your prize lettuces. Certain plants have entered into an unholy alliance with the ants, by providing them with special homes in which the ant colonies can find shelter and a safe place to raise their social hordes. It may not seem to be a good idea to make room for a colony of ants, however social, but it is in part that social behaviour that gains the plant protection. It is the wont of many social insects to guard jealously their own home territory, which means that if some plant-eating insect comes along intent on chewing the myrmecophyte, the resident army of ants will muster for the attack and see it off.

The more you look for interactions the more you find. For instance, the seeds of many plants have a special ant-sized store of food and are, therefore, seized by the foraging ants and carried off to their nests. The seeds that get eaten do not help the plants but the ones that are lost *en route* are thus dispersed, giving the seedling a good start away from the sphere of parental influence.

Not all rain forest plants are so lucky, and for the majority of the trees the real struggle in life is enacted at the time of reproduction. A large juicy fruit may be dispersed by a bird or animal, but for the majority it will be a short, straight drop down to the ground beneath the parent tree. It is then that the problems really begin for it must develop in a habitat that would be completely alien to its parent, a habitat in which there is little or no light and very sparse available nutrients.

It is at this point that the large size of the fruit and seed becomes important, for it represents a store of energy and nutrients, enough to see the seedling well on its way to saplinghood. Once the bulk of the food store has been used up, the growth of the sapling slows down and it then just stays in a state of suspended 'animation', waiting for a place in the high life of the forest.

The chance occurs when one of the canopy trees comes crashing down letting in sufficient light for rapid growth and with the promise of new nutrients in cycle. The sapling that is at that moment in the most advantageous position will probably win through and eventually fill the gap, and all the rest will die, for this is the law of the forest. Every plant, however large or small, must in its lifetime only produce one successful offspring; more than one could be the start of a population explosion, less than that and the population of that particular sort of plant could be on the decline. One from one is the rule and all the time it is obeyed the diversity of forest life will be maintained, a diversity in which every plant and animal in the forest community has a role to perform, and it is always a role well played because failure to do just that may lead to replacement by another more fitted to carry out the job.

'Survival of the fittest' is a catch phrase meant to express the essence of evolution, but where in the diversity of the rain forest with its 3000 different species of trees and its half a million species of insects to say nothing of the other animal groups, is the fittest organism? The answer is all of them; all are fit to play their particular role in a highly diverse, highly successful system, a system in which there is little or nothing for the outsider. And that is exactly what you and your family would be, outsiders in the most affluent society on earth.

One person, Professor Betty Meggers, has spent much of her life trying to understand the relationship of the society of man to the society of the Amazonian rain forest. Part of her study centred on seven indian tribes, all of which appear to have come to terms with different parts of the 5·2 million square kilometres (2 million square miles) of forest.

Two of these tribes, the Tapajós and Omagua, lived on the várzea, which was accessible to the early European explorers. As a result, their cultural pattern had been completely destroyed by 1692. However, records from before this time, and archaeological finds, have allowed a general picture of their way of life to be reconstructed.

All the tribes enjoy much the same style of survival, hunting game and gathering fruit and fish in the numerous lakes, streams and rivers. The natural communities of the forest and the water that drains through it provide them with the bulk of their needs–clothing, transport, building materials, medicine, recreation, food and drink. To exploit these seemingly bountiful resources, they require an immense knowledge not only of what is useful but of how it can be best put to use. They must be botanists, zoologists, orienteers, pharmacists, carpenters, turners, thatchers, joiners, rope makers, potters, armourers, brewers; not jacks but masters of all the trades of life within the forest.

To a greater or lesser extent they all depend on slash-and-burn farming to provide their staple starchy foods. At certain times of the year, which are dictated by the amount of local rainfall or the state of the local river, they destroy an area of forest. The useful timber is put to the manufacture of everything from a new hut to a blow dart, depending on what trees are present in the selected plot, the rest being burned where it falls. This releases the minerals that were locked up in the otherwise useless timber making them available for new growth, provided that the rain does not get them first.

The crops are then planted, a mixture of different crop plants always being used. The advantages of such polycultures are many: they provide a diversity of cover that helps to

protect the soil; a diversity of food that can be harvested over a long period of time; and, in the event of the plot being attacked by disease or marauding insects, there is a better chance that some of the crops will survive.

The main crops are Bitter Manioc (*Manihot utilissima*), Sweet Manioc (*Manihot palmata*), Sweet Potato (*Ipomaea batatas*), yam (*Dioscorea*) and Maize (*Zea mays*). It is no accident that four of these are root crops; they have been selected by man as they are of the greatest use in the forest environment. All four go on producing over a long period of time and the underground parts, which will eventually be eaten, can be left in the ground until required–no deep freeze needed.

Bitter Manioc, which incidentally is the raw material of delicious cassava bread, is of great interest because in its raw state it contains a nasty poison. Before it can be eaten with safety it has to be peeled, grated and the poison soaked out, quite a sophis-

Above Careless ants aid the dispersal of plants. Here seeds collected around an ants' nest are beginning to germinate.
Below Where there are many leaves there are sure to be lots of herbivores (animals that eat leaves). Here a large katydid rests among its lunch. This is just one of the hundreds of thousands of species of insect which live in the rain forests of the world. The wings must help to stop it becoming someone else's dinner, by adding the protection of camouflage.

ticated process. The other staples are much less of a problem. However, like all starchy foods, preparation by cooking and especially by fermentation not only makes them better for eating but also increases their food value.

So the seven tribes employ many of the secrets that make for a successful life in the forest, but each one has evolved its own particular methods. The main and perhaps the most important difference is where they live. Five of the tribes, the Sirionó, Waiwai, Jívaro, Kayapó and Camayurá make their living on the terra firme–that is, the area of high forest that is situated on land above the reach of the rise and fall of the rivers. The other two, the Tapajós and Omagua, made their homes on the várzea, which is the flood plain of the river and its main tributaries and is inundated by floodwaters at least once a year. The floodwaters bring with them soil and silt that have been eroded in part from the vast area of rain forest, but in the main from the hills and mountains (including the high volcanic Andes) which form the catchment of this, the mightiest river in the world.

When I was a very small boy I dreamed of standing at the source of the Amazon and my dream was always set in hot steamy forest. The illusion was shattered when the dream came true and I stood close to one of the river's many sources. The scene was the snow and ice of Mount Cayambe and all around me the products of erosion were joining the trickles of melt that would eventually swell the milky waters of the main river. Every year a new supply of minerals is on the move, and enormous amounts of silt are laid down on the várzea, making it an ideal place on which to raise crops. There is no need for fertilizer bills, it all happens more or less on schedule every year, a fact that was reflected both by the size and the permanency of the settlements of the Tapajós and Omagua.

They had no need to move on. With fish from the Amazon (and there are over 2000 different sorts to choose from, including the much-maligned piranha), the self-fertilizing fields and a little bit of hunting and gathering to add spice to life, there should be no problems. However, this is unfortunately not the case because not only does the river deposit new silt but it can and does wash the old silt away. This unreal estate factor may well benefit another tribe down river but if you have got too many neighbours, life on your bit of 'terra unfirma' could become a problem.

The five tribes of the higher, drier and more stable lands are faced by a much more permanent problem. While under forest their land stays exactly where it is and nothing is added to it except pure rain water. One hectare of high forest only carries a certain amount of game and bears a certain amount of edible fruit, all of which are not only dependent on but are part of the mineral cycle of the forest system. The further you have to walk each day in search of the food you need, the more energy you use up and, as food and energy are one and the same thing, the daily budget, energy obtained to energy used in obtaining it, could begin to get in the red. So, at some stage in the proceedings, it is best to move on to pastures new. In the same way, no matter how good the slash-and-burn crops may be they can not protect the soil as effectively as the forest, nutrients will be lost, the crops will get poorer and the farmer/hunter will have to move on.

The Amazon basin

The mouth of the Amazon, 322 kilometres (200 miles) wide, disgorges a fifth of all the water carried by the rivers of the world. It is, however, not only the mightiest of all rivers, it is also the most back-to-front. River valleys are usually narrowest near the source and widest at the mouth, but the Amazon appears to be built exactly the opposite way round. This is because its direction of flow was reversed by the meeting of two plates (see p52). This back-to-front basin is the home of many thousands of different sorts of animals, all of which make their livelihoods within the bounty of the forests. The map shows the home grounds of the seven local tribes which formed the subject of Betty Meggers detailed study.

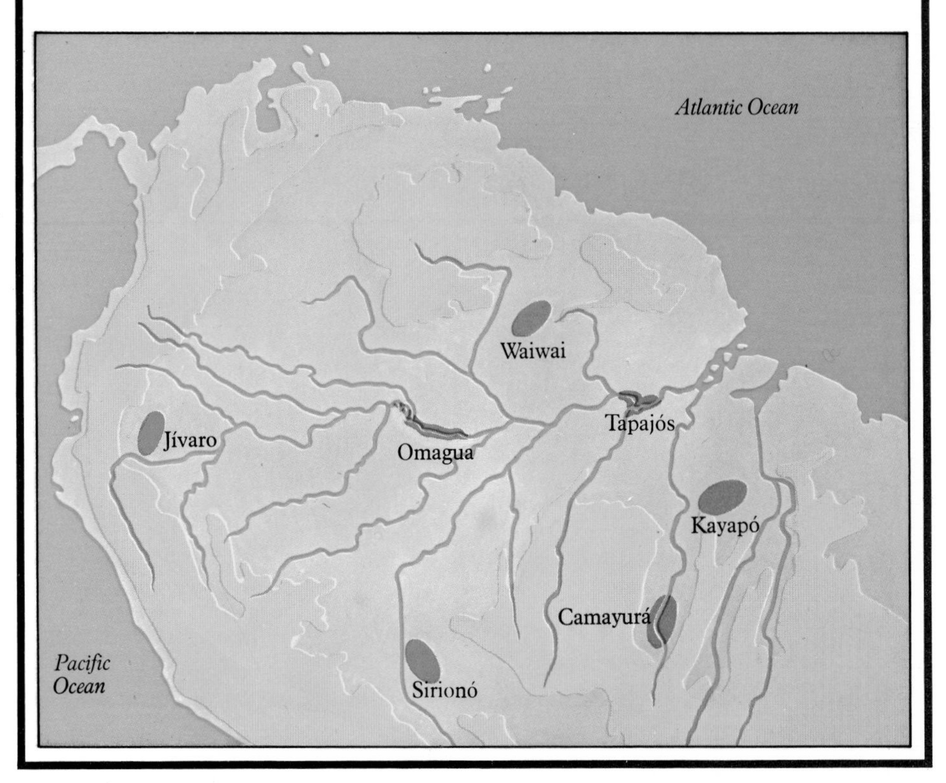

These facts of life on the terra firme are again reflected in the size and permanence of the settlements of the five tribes, although whether they are determined by the fertility of a particular area of forest alone or by the history of the tribe is not certain. What is certain is that the longer a tribe can obtain all its needs from one area of land, the more settled and the more complex its society can become (take a thoughtful look at the table, opposite below).

The Sirionó, who move on at intervals of not more than six months, do very little and have no possessions which are not directly related to the necessities of life–the problems of protection and of obtaining food and drink. Their evolved pattern of life appears to have no time or place for social organization, art, trade or religion. In comparison the Camayurá, who stay in one place for many years, are much more like the tribes of the várzea in that they spend much more of their time and effort in doing and making things which have no direct survival value.

There is one other practice that all these tribes have in common and it is one which is a cornerstone of their success. They all practise population control, the methods ranging from abstinence from intercourse for certain periods, contraception, killing certain unwanted children, abandonment of ill, aged or infirm, to warfare, which in some cases includes head-hunting. Why and how these practices evolved or have been retained from earlier phases of social evolution is a matter of conjecture. It is, however, perfectly clear that they help to check the growth of the population, keeping it in some sort of balance and within the resources.

Too many people, even living at subsistence level, must in time lead to a breakdown of the natural system and eventually to the destruction of the forest itself. There are limitations even on the equator where it rains almost every day and it appears that the seven tribes had evolved to live within these limitations. To put it in Betty Meggers own words 'man and culture in a counterfeit paradise'.

So, how do you reckon you would get on? The place, a nice large plot on the várzea. The time, just as the flood waters are beginning to fall! Best of luck, and in the short term luck would play a very large part in your success, but in the long term adaptability and recognition of those limitations would determine your survival.

Perhaps you would rather shun the tropics and decide to live off the land in your own patch where you know the ropes, but remember the rules of the test 'you can have nothing except the skins you stand up in' and remember that the rules of natural limitation apply everywhere and to everyone, for like the seven tribes of Betty Meggers you are part of the process of evolution.

Above An Indian of the Camayurá tribe.

Below Cultural traits among the seven tribes.

(●? = Probably present.)

Traits	Sirionó	Waiwai	Jívaro	Kayapó	Camayurá
SEDENTISM					
Village population	± 80	± 25	± 40	150+	± 110
Village permanency	under 6 mo.	± 5 yrs.	± 6 yrs.	Indef.	± 10 yrs.
SOCIAL ORGANIZATION					
Household chief	●	●	●	●	●
Multihousehold chief					
Temporary (war)			●		●
Permanent				●	●
Village council				●	●
Non-kinship based associations				●	
PART-TIME OCCUPATIONAL SPECIALISTS					
Shaman		●	●	●	●
Arts and crafts					●
TRADE					
Within village					●
Between villages or intertribal		●	●		●
Formal market					●
SPECIALIZED STRUCTURES					
Men's house				●	
Flute house					●
Chief's house				●	
RELIGION					
Ceremonial posts (idols?)					●

Traits	Omagua	Tapajós
SEDENTISM		
Village population	300+	300–2 500
Village permanency	Indefinite	Indefinite
SOCIAL ORGANIZATION		
Household chief	●	●
Multihousehold chief		
Village	●	●
Multivillage	●	●
SOCIAL STRATIFICATION		
High Chief	●	●
Slaves	●	●
FULL-TIME OCCUPATIONAL SPECIALISTS		
High chief	●	●
Shaman	●?	●?
Arts and crafts		●?
TRADE		
Within village	●?	●?
Between villages or intertribal	●	●
Formal market		
SPECIALIZED STRUCTURES		
Temple or shrine	●	●
Storehouse or granary	●	●
Chief's house	●?	●?
RELIGION		
Idols	●	●
Prayers and offerings		●

Chapter 2

The crucible of life

To understand these limitations, we look at the origin of life itself and find that the laws which govern the workings of the universe at all levels, from atoms to galaxies, dictate that life chemicals based on the element carbon had to happen. Life had to evolve.

In theory

What are little boys made of?–slugs and snails and puppy dogs' tails, that's what little boys are made of.
What are little girls made of?–sugar and spice and all things nice, that's what little girls are made of.

ANON

I do not know whether Anon was a chauvinist sow or not, but the important thing is that all the items in the jingle are made from the same things. They are all made of organic chemicals which are, in turn, made of elements that once formed part of the earth's crust, its oceans or its atmosphere.

The earth has a mass of $5{\cdot}97 \times 10^{24}$ kilograms ($13{\cdot}13 \times 10^{24}$ pounds), a big number and one that really matters because that is all the matter we have got. The outer shell of the earth, crust, sea and atmosphere, is made up of ninety-two elements, which are by definition substances which cannot be broken down into anything simpler by a chemical process. Therefore, if anything exists on the surface of the planet earth it must have either come from outer space or it must be made of one or more of these elements. An intriguing fact is that of the ninety-two elements only twenty-two of them are found as components of naturally occurring organic chemicals. Of these carbon (C), hydrogen (H), nitrogen (N), oxygen (O), phosphorus (P), and sulphur (S), are the most important in terms of abundance and function and the first four are the most important of all. Why is it that the chemistry of life is so conservative in its make-up? To be able to answer this question it is necessary to know something about the structure of matter.

All matter is composed of atoms. Hydrogen is the lightest and uranium the heaviest of the naturally occurring atoms. The diagram (right) shows the structure of the ten lightest elements, each of which looks not unlike a miniature solar system. The central nucleus is made up of two types of particle called neutrons (no charge) and protons (positively charged) and together they give the real weight to the whole system. Around the nucleus tiny, effectively weightless, satellite par-

The elements of the earth and the life it now supports

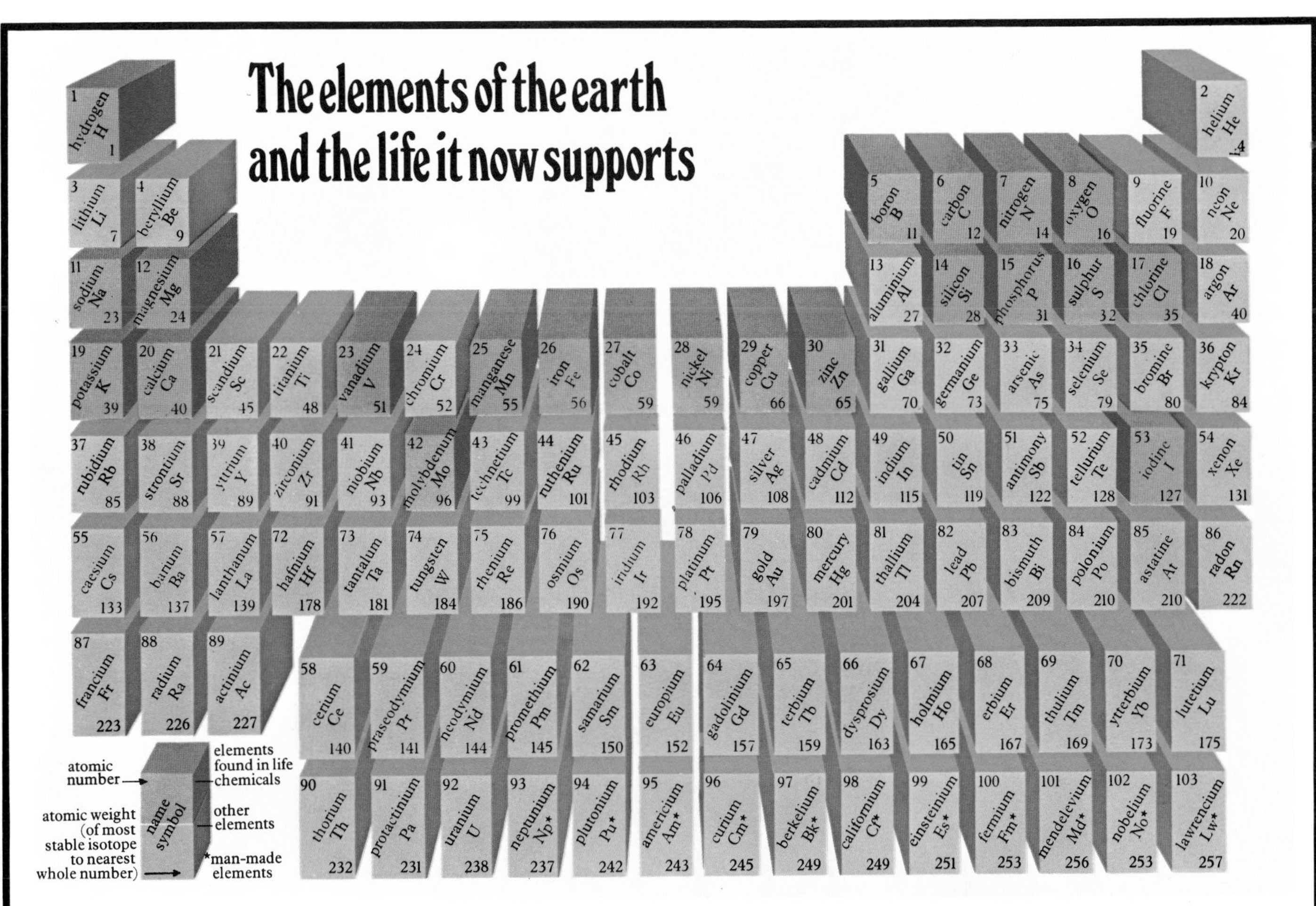

Above The Russian chemist Mendeléeff, in 1869, first saw the order in the then known elements. His periodic table, which arranged the elements in ascending atomic weight, has stood the test of time including the discovery of many new elements. It not only brought together all the groups of elements which had similar properties but also allowed Mendeléeff to accurately predict the properties of the new elements before they were discovered.

This is a special edition of the table which highlights all the elements that play some role within the chemistry of life. It is of interest that they occur in groups on the table – carbon, nitrogen and oxygen, three of the 'backbone elements'; manganese, iron, copper and zinc, which are essential micronutrients for the healthy growth of plants; and sodium, potassium and calcium, which are important in the regulation of certain properties of blood. Perhaps as we learn more about the ordered structures of the biochemicals we will begin to understand more about how the properties of these twenty-two elements fitted them to the roles they play in the chemistry of life.

Below The first ten elements in the periodic table in ascending order of weight and complexity.

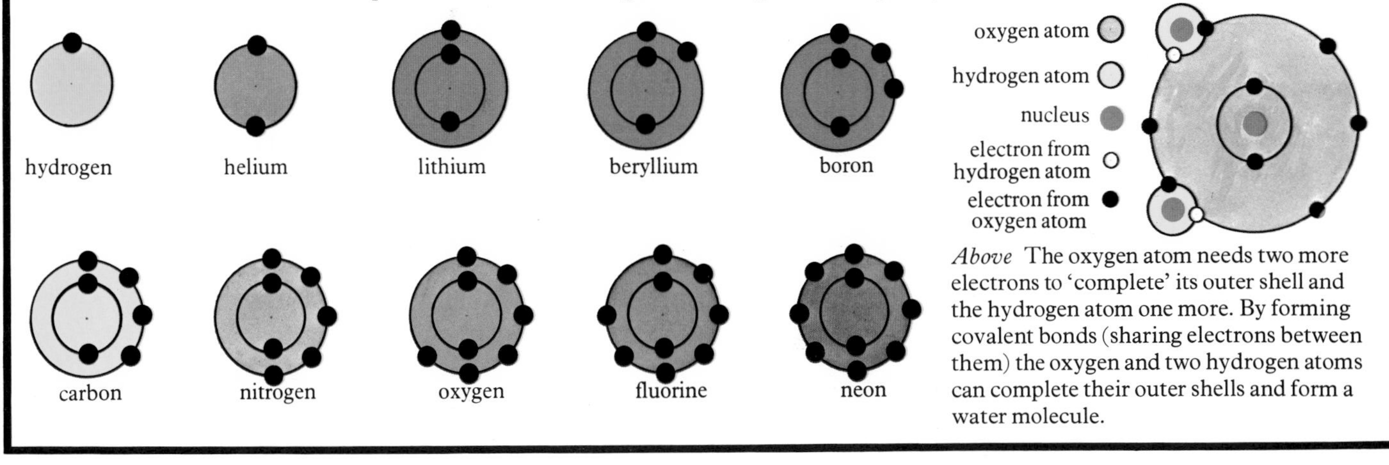

Above The oxygen atom needs two more electrons to 'complete' its outer shell and the hydrogen atom one more. By forming covalent bonds (sharing electrons between them) the oxygen and two hydrogen atoms can complete their outer shells and form a water molecule.

ticles called electrons (negatively charged) move around fixed orbits. The number of electrons always equals the number of protons, so that the whole atom is electrically neutral. It is the number and position of the three types of particles in the atom that makes each element slightly different from all others. It must be noted that the planetary diagrams are simply a nice shorthand way for us to visualize the structure of the atom, no more.

If you look carefully at the ten lightest atoms you will see that two of them, helium (He) and neon (N), look nice and symmetrical, almost self-satisfied, the electrons in their orbits being nicely balanced out, while all the rest seem to have one or more electrons too few or too many. Helium and neon are inert gases, which means that it is very difficult to make them combine with other elements to form compounds. Their property of inertness is locked up in those 'self-satisfied' orbits. That is one

A large-scale demonstration of the properties of the two lightest elements. The helium-filled *Mayflower* (*above*) operates trips over Miami in the winter season. The hydrogen-filled *Hindenburg* (*top*) exploded over Lakehurst, New Jersey in 1937.

reason why helium is used today for filling balloons–remember what happened to the *Hindenburg* which was filled with hydrogen? Helium and hydrogen are the two lightest elements, ideal for filling up airships, but one is inert and the other explosively reactive, combining with oxygen with a very loud bang to produce water (HOH), a compound.

Any element which has incomplete electron shells is reactive. To complete (satisfy) their electron shells a hydrogen atom needs one, an oxygen atom two, a nitrogen atom three and a carbon atom four electrons, and that is why they are all reactive, making them good raw materials for the chemistry of life.

These four elements have an additional property. They have the capability of sharing one or more pairs of electrons so binding themselves to other atoms with single or multiple covalent bonds. As the strength of any covalent bond is inversely proportional to the weight of the atoms it is binding, the covalent bonds between C, N, H and O must be the strongest possible.

Carbon atoms have one final important property–they can combine with each other by accepting or donating four electrons. In fact, four carbon atoms can share electrons in a covalent structure which has great versatility, especially when it comes to form and function.

If you want proof, look in the mirror. That is diverse, brilliant you, and the bulk of it is made of proteins, fatty acids and carbohydrates with a covalent carbon at the core of each.

So it was that the structure of these four elements fitted them perfectly to the central role that they play in the chemistry of life. They were the fittest and have survived throughout evolutionary time and today play the same basic role that they played at the dawn of evolution more than 3 200 million years ago.

The burning question is how did it all start? Here opinion is divided in a multitude of directions. At the two extremes of one line of argument are the opinions that a superbeing, best called god, 'pushed the button' creating life, and that it just happened. It is well to remember that it is not just Christianity which teaches the former. Most religions of the world have a basic concept of god the creator, and not all scientists are atheistic in their outlook. The concept of a superforce causing creation to happen is obviously better than a wishy-washy 'it just happened'. At least the gods of any religion are worthy and often very resilient Aunt Sallys–hypotheses to be proven or swept aside.

So what are the facts? We know that our bodies are made of no more than twenty-two elements, none of which, by themselves or in combinations other than those based on covalent carbon, can be said to be alive. We know that there are more than one million other sorts of living organisms on earth and that they are all made of combinations of some of the same twenty-two elements, but as far as we know we are the only organism that has the ability to worry about such things. That ability is best called consciousness, and that is possibly the one major attribute that sets man aside from all other products of evolution.

You may beg to differ on this point, be my guest. I am willing to concede that your pet pussy cat may worry about where his next meal comes from, but I do not think that he worries about being a pussy cat. Now that fact must make us feel pretty important beings and might lead us to conclude that 'super god' created 'super man' in his own image, endowing him with consciousness which made him aware of the fact. It is perhaps not quite as exciting to conclude that we are no more than part of a chemical process which had to happen because of the natural laws that govern the universe and all its parts.

The planetary diagrams of the atoms may well be misleading as far as pure physics is concerned, but they do firmly remind us that all matter, no matter how large or small, simple or complex, obeys the same rules. There is no opting out clause. It is my own considered opinion that it is a much grander view of creation to envisage a god creating the rules that govern the universe and made it all happen than to think that god created the myriad of insects and other organisms one by one.

What then are the rules, what are the hypotheses and how do they stand up to the facts?

Rule 1 *Matter/energy* (these are different forms of the same thing) *can neither be created nor destroyed, but can only be changed from one form into another.*

Rule 2 *When energy is changed into another form, some of it is always degraded, that is, in a less easy-to-use state.*

Rule 3 *All systems tend to degrade the energy that is available to them.*

Rule 4 (This is, perhaps, a proposition, rather than a rule.) *Time is a function of change because if nothing changed then there would be nothing to mark its progress.*

Somewhere at sometime in the past, carbon, oxygen, hydrogen and nitrogen came together in the right proportions and with sufficient energy to 'weld' the covalent bonds, producing chemicals which had certain basic attributes of life.

It is, in fact, possible to make it happen under experimental conditions. Take four gases hydrogen (H), ammonia (NH_3), methane (CH_4) and water (HOH) vapour and subject them to a miniature electric storm in a closed container. In time, four

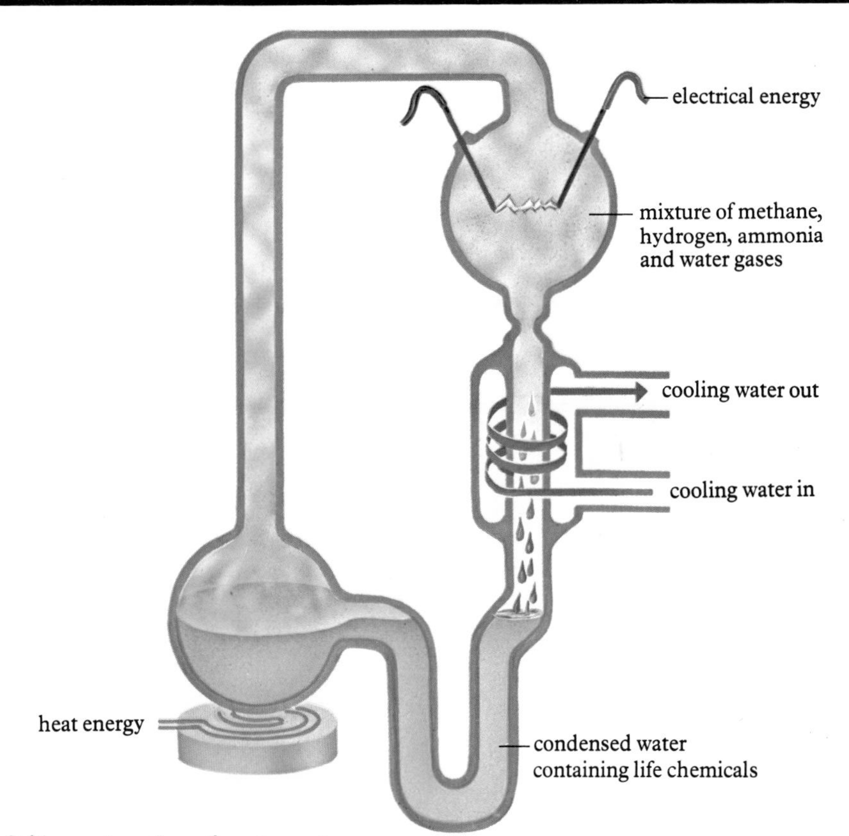

The vital mix *in vitro*

(*In vitro* means in an artificial environment.)

A closed-circuit environment, not unlike the one in the diagram, was first constructed by two American scientists called Miller and Urey. Its main constituents were the silicon of the glassware, the metal of the electrodes, the gases methane (CH_4), ammonia (NH_3), and water (HOH), all of which were present in the environment of the primitive earth. They supplied the environment with two external sources of energy, heat (cf volcanic activity) and electrical discharge (cf thunder storms) both of which were of common occurrence on the surface of the primitive earth. In time, new chemicals appeared in the flask. These included four amino acids, some fatty acids and urea, all of which deserve the title of 'life chemicals'.

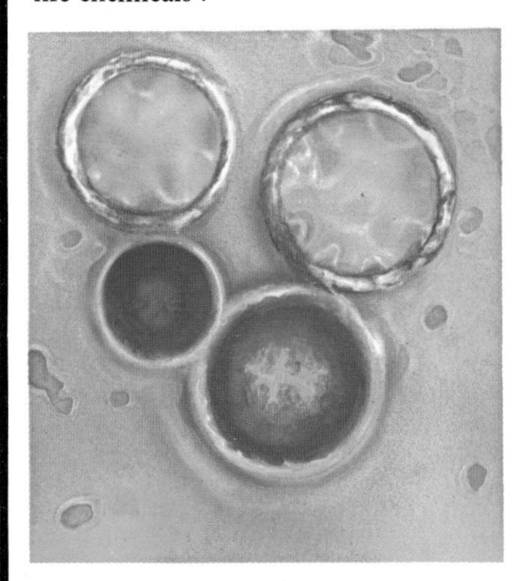

Protein spheroids

These tiny spheres were produced by heating a test tube containing a suspension of protein molecules in water. They are not alive and although they lack complex membranes and internal structures they are of similar shape and size to real live cells.

The vital mix *in vivo* *Overleaf*

(*In vivo* means in a living organism.)

Earth, fire, air and water – the early philosophers believed that these were the four elements of which the world and everything on it were composed. Although we now know that the planet and its atmosphere are made up of ninety-two chemical elements we still believe that it was the basic properties of these elements and the rules which govern their existence which set in motion a process which would in time cover the earth in a mantle of living green.

amino acids, some fatty acids and urea can all be detected in the container; none of them were present before and all of them are life chemicals. This experiment has been successfully repeated on a number of occasions and more recently it has been extended, using a mini-volcano as a source of energy, to a stage where tiny spheres of protein, each enclosed in an envelope, have been formed. Of course, this does not mean that life has been created *de novo* but, since all these gases and conditions (thunder storms, volcanoes and so on) have always existed on the earth, the results of the experiment are of more than passing interest.

We know that there was no element of chance in the outcome of the experiment. The properties of the four elements listed above dictate that, given the right conditions, covalent carbon structures will be formed. In a nut shell, the presence of life chemicals on the face of the earth is really no more surprising than the presence of the rocks themselves, given the right conditions *it had to happen* and you really can not expect a much more positive affirmation of the act of creation than that.

Now for the concrete evidence! If we look back through the record in the rocks we are presented with a number of firm dates after each of which we have fossil evidence of the existence of various forms of life and, as time progresses, the life forms get both more diverse and more complex.

Four thousand six hundred million years ago, $5{\cdot}97 \times 10^{24}$ kilograms of matter took up its orbit in the solar system to become the planet earth. The rocks that were laid down over the next 1200 million years contain no fossils although there is evidence in the younger ones of the presence of covalent carbon compounds including the building blocks of such complex molecules as chlorophyll.

About 3400 million years ago we have real fossils, good hard evidence of the presence of very simple cells which look not unlike the bacteria and blue-green algae which we find on the earth today and we call prokaryotes. Roughly 1500 million years ago, more elaborate cells with a complex internal structure are found.

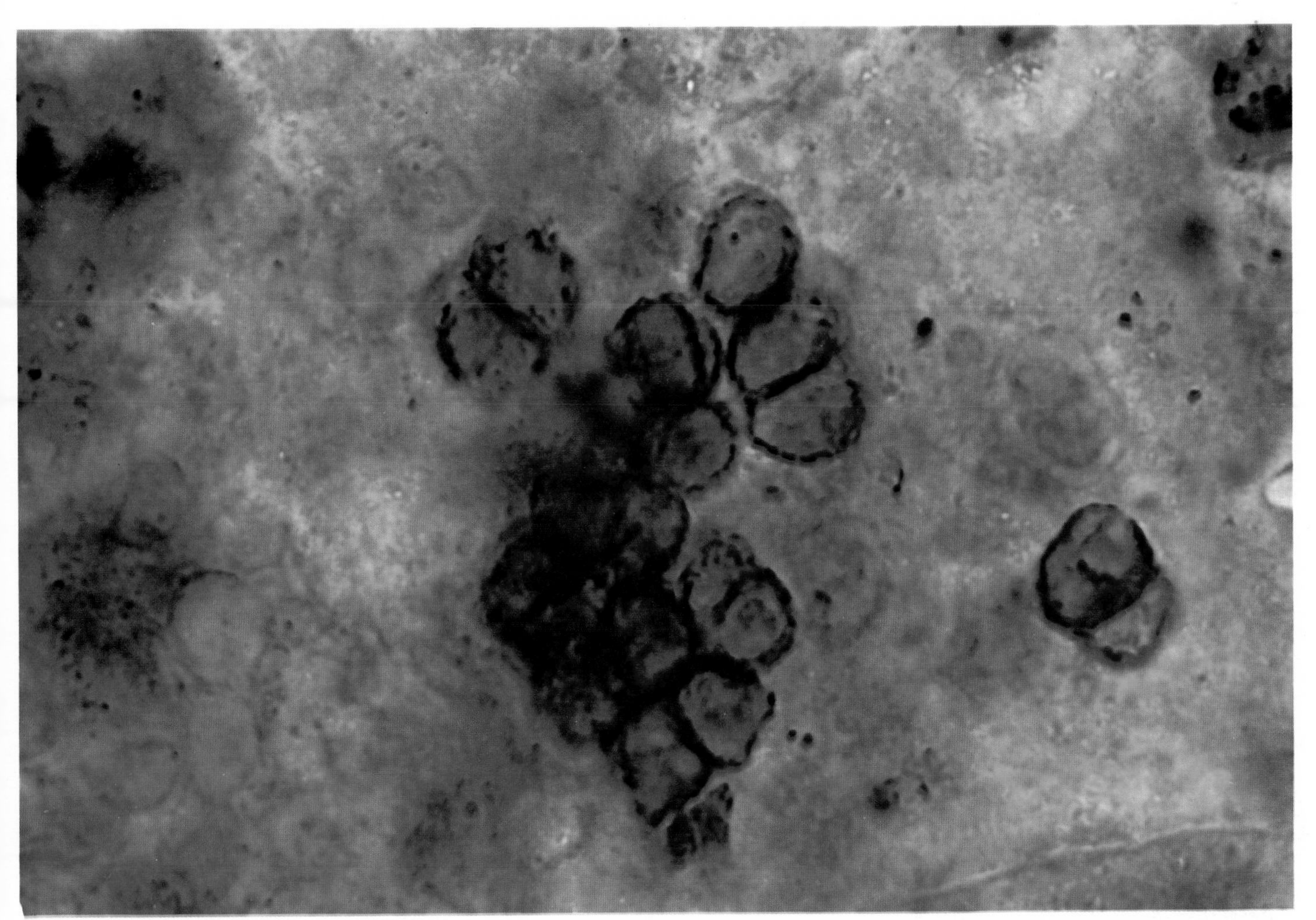

Prokaryotic cells dividing. These were found near the base of the Amelia Dolomite, McArthur Group, Northern Territory, Australia, and are approximately 1600 million years old.

The complex internal structure of organelles are indicative of differentiation of labour, different parts of the cell doing different jobs, as in the modern organisms which we call eukaryotes. Towards the end of this period of time complex aggregates of cells are found.

Animal fossils are found for the first time 600 million years ago. All have soft bodies and they include assorted worms, jellyfish, sea-pens and echinoderms, all of which have modern day counterparts, and a number of types which have no modern relatives.

The really big shock occurs 30 million years later. From that point on certain rocks are full of fossils, a fact that makes them so easy to distinguish from the older strata in which fossils are very few and far between that they are given a new name, Cambrian (named after Wales, where rocks of this type were first described). This geological period is packed full of life and the surprising thing is that there are not only lots of fossils, but there are lots of different sorts of fossils present.

This sudden appearance of a whole cross-section of animal life, the majority with hard readily fossilizable skeletons, has caused an awful lot of heartache. Are the special creationists right? Did god decide around 570 million years ago 'Let there be fossilized animals'?

The real question is easier to answer, for it is simply how quick is 'sudden'? If it all happened in six days then the whole science of geology is a figment of a lot of very brilliant peoples' imaginations and the fossil evidence is the biggest red herring ever created (or should it be evolved?). In comparison, 30 million years, for that is the length of the early Cambrian during which the fossil evidence gradually builds up, is surely enough. So, the evolutionary 'explosion' in the early Cambrian is not proof against evolution, but is among the strongest pieces of evidence to support it. It emphasizes the importance of time.

Whenever experiments aimed at producing life chemicals have been carried out successfully one important ingredient has been time. Even though the reactions are predictable from the properties of their raw materials it took time to get the conditions just right, even in the confines of a small flask. Even if, as we learn more and more about the structure and properties of the complex chemicals which form the core of the life processes, we find that they too had to happen, it was going to take an immense amount of time to get each set of conditions just right and on a worldwide scale. The evidence, more of which is at this moment being extracted from the rocks, indicates that it took over 4000 million years to get all the formulae right, ready for a world take-over bid. To put it in perspective, it is worth surmising on the probable problems and the main breakthroughs.

Phase 1 Background to evolution
Covalent carbon chains are being formed at various places on the earth where the conditions are right. Some of these are amino acids, the building blocks of proteins, others are simple

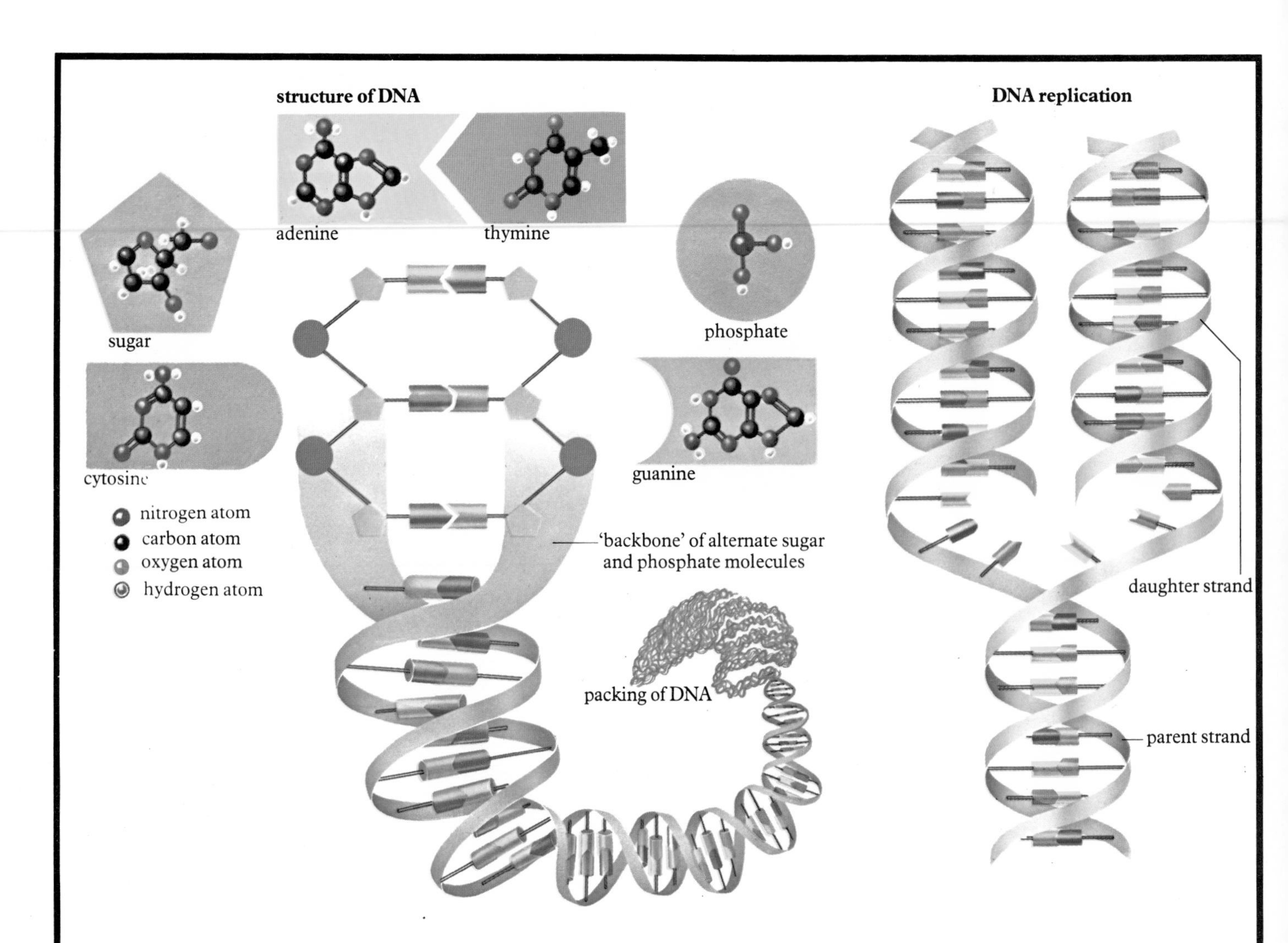

Inside information

Inside the cells of every living organism there is a library of chemical information stored on giant molecules (macromolecules) called nucleic acids. The molecule in the diagram is called deoxyribose nucleic acid (DNA for short) and magnified up it looks not unlike a twisted rope ladder, each rung of which is made up of chemicals called 'bases'. There are four different bases found in each DNA molecule and they are called adenine, thymine, cytosine and guanine. Note that these four bases are composed of just carbon, hydrogen, nitrogen and oxygen. Adenine must always pair up with thymine and cytosine must always pair up with guanine. The order of the bases down one side of the ladder will therefore determine the order of their partners down the other. Imagine that the two halves of the ladder were pulled apart and each half was sent away to a different craftsman whose job was to make a new ladder. It is easy to see that following the basic rules set out above each would produce a ladder which was identical in every way to the original one. What an ideal way of storing and duplicating exact information, and that is in fact the function of the complex library steps in each living cell.

There are millions and millions of rungs on each DNA ladder and it has been discovered that sections of the ladder, which may consist of between a few hundred to a few thousand twists, act as a code which spells out the detailed instructions of the specific proteins that go to make up each living cell. Such a segment has been called a 'gene'.

So it has been throughout evolutionary time, the vital information being passed on from generation to generation. The oldest code in the world has held the secret of life from its very beginnings, a secret which all cells share in the core life functions they perform. It has also been the substance of evolution, an ever-expanding library of evolutionary change recording for posterity the ever-increasing potential of the living chemicals.

The code remained a secret at the core of life until 1962 when the brilliant, deductive mind of *Homo spaiens* in the guise of Maurice Wilkins, James Watson and Francis Crick announced the cracking of the code. I do not know how many 'genes' it was that gave them this ability. All I hope is that I have got some like them to pass on to my offspring.

fats which have the basic properties to make membranes, and some are simple sugars structured for energy storage and precursors of storage and structural molecules like starch, glycogen and cellulose.

Phase 2 Coacervates, first forms of life?

In certain places on the earth, possibly in tidal pools near the edge of the sea where concentration took place at regular intervals, the products of phase 1 were moulded into aggregates of various shapes and sizes. The most likely structures are spheres of protein, each bounded by a membrane which effectively shielded its contents from the diluting effects of the outside world. Such a boundary layer would lead to control of the internal environment, perhaps making it a fitter place in which chemical evolution could happen faster. Such structures are best called coacervates.

Phase 3 Inside information

Just as conditions existed on various parts of the earth which caused the formation of covalent carbons and coacervates, so too conditions must have existed which would cause their destruction. The fact that chemical evolution progressed is tantamount to proof of a positive directive force in evolution; that the 'had to happens' were less of a matter of chance than the 'destructions'. Nevertheless, the fact that there was the possibility of destructions must mean that on many occasions new and novel structures, including major advances in the chemistry of life, could have been wiped out. What was required was a chemical mechanism with the property of storing and faithfully reproducing the novel chemical information.

The fact that all contemporary organisms contain substances called nucleic acids which can do just that, and that in all cases the method of coding and process of replication of the information is the same is proof enough that once this step had been taken, the chemistry of life recognized that it was on to a good thing. How that step was taken we can not even hazard a guess at, but once a coacervate was endowed with such a mechanism it was assured of success, so much so that at this stage it must be called an organism.

Just think of the advantages such a system would confer on the organism endowed with it. Once it had happened it had to happen again and again and again, an information explosion, with more and more of the same covalent carbon compounds and more of the same organisms. An organized body, organized to produce more of itself–what could hold it back? The answer is energy and raw materials. The first organisms, bristling with an armoury of information could go around coalescing with the lesser coacervates, appropriating their energy and information for their own ends.

If nothing further had happened the living world of original organisms would have remained at a low ebb dependent upon the rate of the *de novo* synthesis of life chemicals, for that was all the energy that was then available.

Phase 4 The green light

Again there was a need, this time for a transformer, a substance that could trap the energy of the sun and a process that could make it available to the life processes of the organisms, and again it happened. We know that the substance that filled the role was chlorophyll b. This green pigment, initiating a process we now call photosynthesis, put the go in evolution and still today drives over 99·9 per cent of the living world.

Remember rule 3 which stated that any system of matter will tend to degrade the energy available to it. If that system of matter has the ability to use some of the energy to make more of itself, then evolution is on to a winner.

Success compounds with success. The main by-product of photosynthesis was free oxygen, a substance which until that time had never existed in its free state on the surface of the earth. (As oxygen is so reactive, free oxygen can only exist where it is being continuously produced.) Proof of this fact is written all over the red beds, the earliest of which are less than 2000 million years old. Their colour is due to oxidized (ferric) iron and so they indicate the presence of free oxygen.

The more photosynthesis there was, the more free oxygen was released into the atmosphere, and this had two effects of enormous importance. It made possible the ozonosphere, a layer in the earth's outer atmosphere that is rich in ozone (O_3), a gas which effectively absorbs ultraviolet radiation. The continuing process of evolution was thus screened from the most harmful part of the sun's radiation (remember your sunburn).

It also made the release of energy by the process of respiration much more efficient. In the absence of oxygen the process of respiration is limited to anaerobic fermentation which may make good alcohol but it does not release much energy from the sugar stores. In the presence of oxygen, respiration can release from bondage thirty times as much energy, and the by-products carbon dioxide and water, both of which are the raw materials for photosynthesis.

No wonder that all of a sudden (in evolutionary terms) the Cambrian fauna sprang into abundance. The conditions were right and it had to happen. If you are worried by all the ifs, buts and perhaps's, so am I. All I can suggest is that you find another explanation for the record in the rocks.

There were 4000 million years of preparation, 30 million years of adaptation, we have only been here for 100000 years and we are already worrying about it. We cannot begin to hazard a guess at how our consciousness evolved, but we can, I believe, ask a pertinent question–does it fill a need? That is the subject of this book.

In practice

It is easy to write about the origins of life, but how do you visualize it? In a world that is overflowing with life, where do you go to see its origins? What we needed was a location that was devoid of life, yet had the potential for it. From the results of the experiments of Mariner 2, the surface of Mars appeared to be the perfect setting, but I am afraid it was, to put it in business jargon, beyond budget. We, therefore, settled for the fact that we were earthbound, and that we were looking at least 3000 million years too late, and started looking at maps of the world.

What we required was an area with lots of sunlight, water, volcanic activity, and yet, above all, it must look like a desert. There is such a place

The geological time-scale

The chart on the left shows some of the major events which shaped our planet for living things, from the formation of the earth to the present day. The chart on the right gives a more detailed history of the last 570 million years, particularly the evolution of life. The dates given are those which are accepted at present, but they are subject to revision when new finds are made or more accurate methods of dating rocks are discovered.

The bulk of our information about past life comes from fossils. The first animal fossils occur in rocks which are 600 million years old. Fossils are common in Cambrian rocks and from this time until the present day, the evolution of life is documented by a fairly complete fossil record. They have also contributed much to our knowledge of continental drift and past climates.

Fossils are the remains of plants and animals preserved in rocks such as limestone, sandstone, shale and chalk, which are all sedimentary rocks. Aquatic sediments present better conditions for fossilization and so aquatic life is much better represented in the fossil record than land animals and plants.

Fossilization Only a tiny fraction of the life that has existed has been preserved to be examined by us today, because several conditions are necessary before fossilization can take place. The organism is much more likely to be fossilized if it has hard parts such as a shell, test or bones and these must be buried rapidly if they are to escape break-down or decay. In addition, the sediment must be suitable for fossilization and must have remained intact up to the present day!

Fossils may be the original hard parts of the animal but usually they have been changed in some way. Minerals from the surrounding rock may enter, increasing the hardness and weight of the specimen. This process is known as petrifaction. Alternatively, the remains may be dissolved away and replaced by other minerals (replacement fossils). If this happens gradually, the internal structure of the fossil may be preserved, but if the remains are dissolved away rapidly, a mould or cast will result and the internal structure will have been lost. Sometimes the organism may be converted into a film of carbon, a process known as carbonization.

Animals leave additional clues such as footprints, tracks, trails, burrows and droppings which, if preserved, are known as trace fossils.

The classification and correlation of these fossil clues together with the dating of rocks by the thickness of sedimentary deposits, the relative amounts of radioactive isotopes and their products, or indeed by the fossils themselves, have enabled a history of life to be pieced together, a history which is being added to all the time.

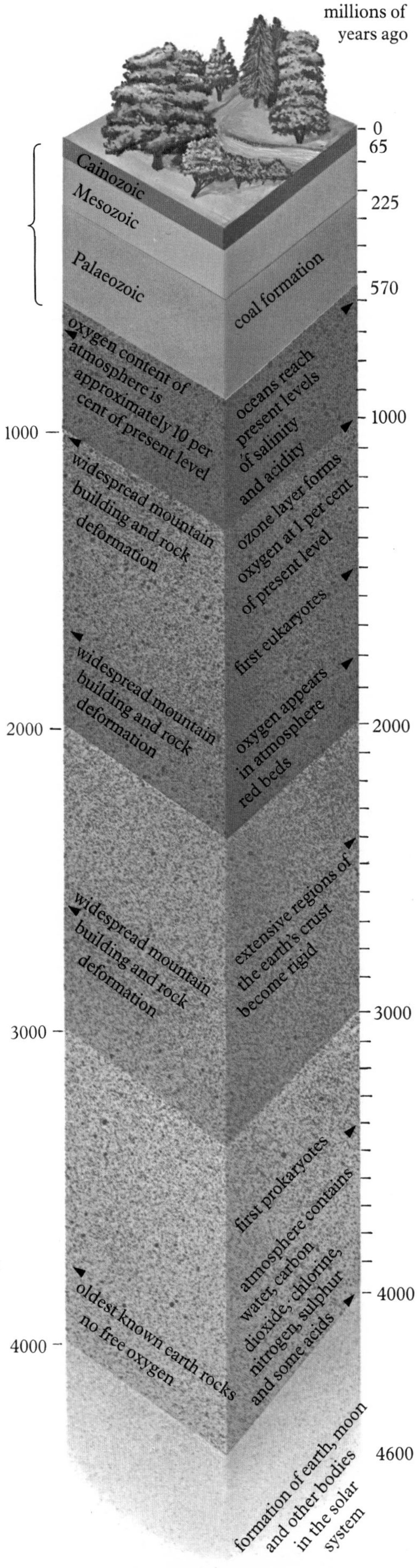

Abundant fossil faunas during this period of time allow quite a detailed reconstruction.

The first remains of life forms preserved as fossils in rocks are of prokaryotic organisms. As these life forms slowly evolve, oxygen generation commences. Oxygen appears in the atmosphere from the oxygen-releasing synthesizers, the blue-green algae, and starts the slow process of changing a carbon-dioxide-rich atmosphere to an oxygen-rich atmosphere. Sedimentary rocks are laid down which contain large amounts of oxidized iron. These are known as red beds because of their rusty colouring.

The lack of an ozone layer meant that a high level of ultraviolet radiation reached the surface of the earth. In certain circumstances this radiation, together with the gases of the atmosphere and condensed liquids on the surface, could produce organic compounds, such as amino acids.

Many important rock-forming events occurred, three of the most widespread being indicated on the chart.

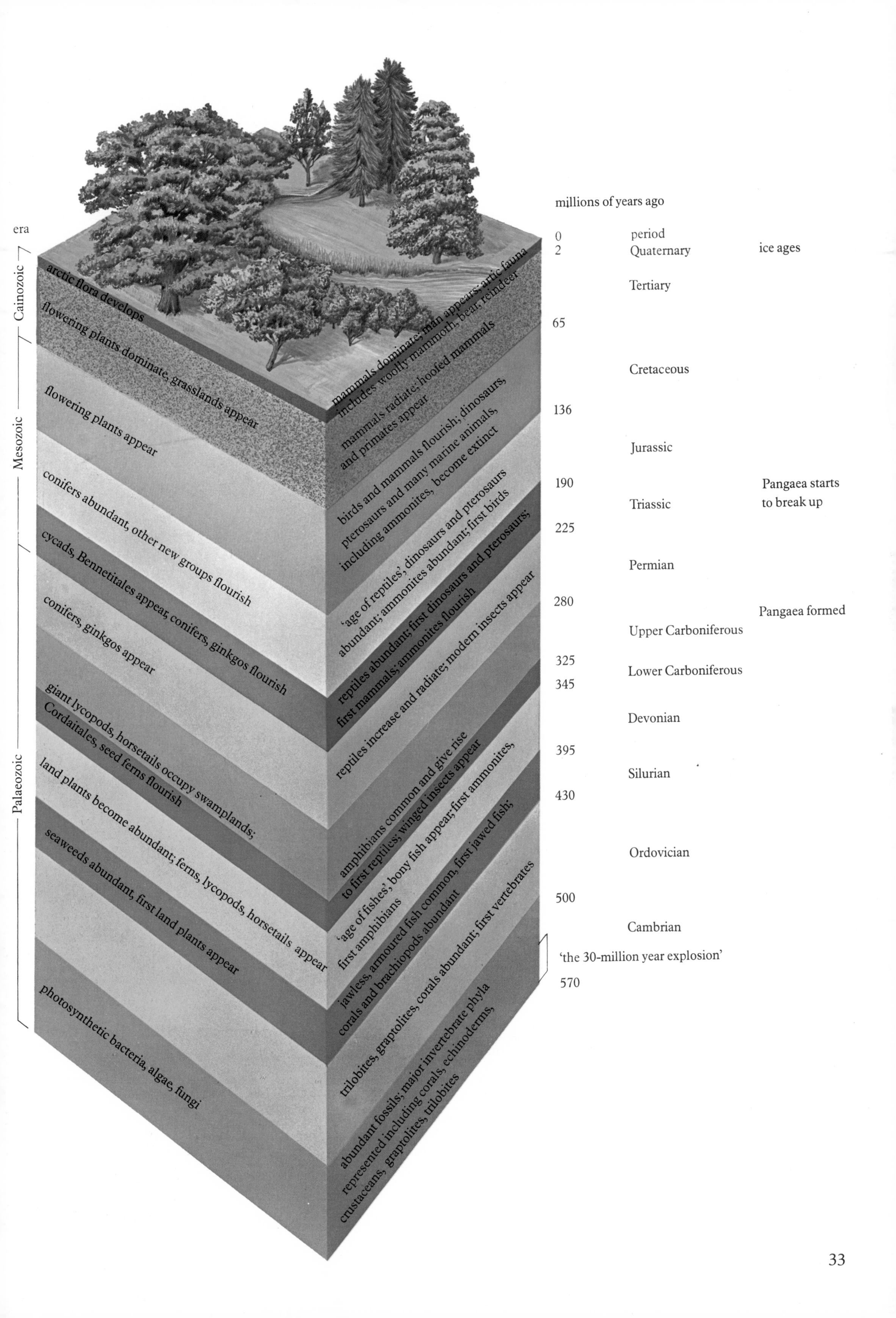
millions of years ago
era
Cainozoic
Mesozoic
Palaeozoic
0
2
period
Quaternary
ice ages
Tertiary
65
Cretaceous
136
Jurassic
190
Pangaea starts to break up
Triassic
225
Permian
280
Pangaea formed
Upper Carboniferous
325
Lower Carboniferous
345
Devonian
395
Silurian
430
Ordovician
500
Cambrian
'the 30-million year explosion'
570
arctic flora develops
flowering plants dominate, grasslands appear
flowering plants appear
conifers abundant, other new groups flourish
cycads, Bennetitales appear, conifers, ginkgos flourish
conifers, ginkgos appear
giant lycopods, horsetails occupy swamplands; Cordaitales, seed ferns flourish
land plants become abundant; ferns, lycopods, horsetails appear
seaweeds abundant, first land plants appear
photosynthetic bacteria, algae, fungi
mammals dominate, man appears; artic fauna includes woolly mammoth, bear, reindeer
mammals radiate; hoofed mammals and primates appear
birds and mammals flourish; dinosaurs, pterosaurs and many marine animals, including ammonites, become extinct
'age of reptiles'; dinosaurs and pterosaurs abundant; ammonites abundant; first birds
reptiles abundant; first dinosaurs and pterosaurs; first mammals; ammonites flourish
reptiles increase and radiate; modern insects appear
amphibians common and give rise to first reptiles; winged insects appear
'age of fishes', bony fish appear; first ammonites, first amphibians
jawless, armoured fish common, first jawed fish; corals and brachiopods abundant
trilobites, graptolites, corals abundant; first vertebrates
abundant fossils; major invertebrate phyla represented including corals, echinoderms, crustaceans, graptolites, trilobites

way down at the bottom of the Red Sea–the Afar Triangle, a tract of hot desert that spans the frontier of Eritrea and the city of Djibouti. It is reputed to be one of the sunniest places in the world, ideal for evolution (at least as regards light energy), it is pockmarked with hot springs and other effects of recent volcanism, and there are two lakes, Assal and Abbe.

So, after much reading and research, we went to investigate one of the weirdest and most hostile environments on earth. Heading for Lake Assal it all looked perfect, but where was the water? Picking our way through a maze of lava flows and heat-shimmer mirages, the four-wheel-drive convoy made its way to the spot marked 'lake' on the map. There it was, a sheet of water surrounded by a gleaming white apron from which the albedo was so strong that it was impossible to proceed without dark sunglasses.

The wide, white apron was salt, a thick layer which blanketed the terrain all around. The temptation for a swim was too great, and in we went, or rather in we tried to go. The first necessity was a good, strong pair of boots, because the floor of the lake was made up of millions of needle-like salt crystals, each one honed to perfection by crystallization. The second was a lot of splashing, for although it was easy to float about on the top of the buoyant brine, it was not such an easy or pleasant experience to get submerged. In fact, now I know what an upside-down iceberg feels like, five sevenths out of water and the rest below. With a supreme effort, total immersion was possible, but oh, the pain of saturated salt solution on your lips and in your eyes. The experience was, I must confess, cooling, but in no way refreshing, the end result being, at least for me, a Father-Christmas-type beard, rimed with salt.

The reason for all the salt lies in the fact that Lake Assal was once much bigger, an arm of the Red Sea itself. It was cut off by volcanic activity and, since then, it has been doing a disappearing act, the water evaporating into hot, dry air leaving the salt behind. Not an ideal place for a swim, and surely no fit place for life.

The landing was, if anything, more painful than the launching. My feet were protected by boots, my rear end was not so lucky, and the waves not only rolled me over, but also rubbed the salt into the wounds.

I was, however, not the first old salt to be washed up on the shore. Along the strandline were heaps of tiny fish, dead and marinaded where they lay. They showed that there was potential for life further out in the deeper, sweeter water but that once washed up into the shallows where evaporation would have its greatest effect–oh, the pain of it. It seems almost impossible to evaporate a sea, but there all around us was proof and, if the sun can do that, think of the potentials for life. Above the windrows of salted fish were a series of miniature storm beaches, each one made up of spherical pebbles. Closer inspection revealed them to be not pebbles at all but balls of salt, millions and millions of them–the world's first soluble marbles–each one produced by the wind rolling the angular crystals together.

What a place for life–the salts of the earth, the water of the sea, and the gases of the atmosphere. All we needed was the fire of volcanic activity, and the stage would be set.

All around the other lake the products of past volcanism were abundant–lava that looked as if it was still in the process of solidification was heaped up over older flows, splashes of yellow marked the positions of fumaroles, and, here and there, bubbling mud and water warned that activity was not all that far below. There was one problem. Around each hot spring there were zones of lush vegetation and skeins of geese were winging their way in to make short stop-overs in their flight across the desert. Beautiful and interesting as they were, our requirements were for a desert with the potential of life, not an oasis that was already overflowing with it.

We wended our way back through

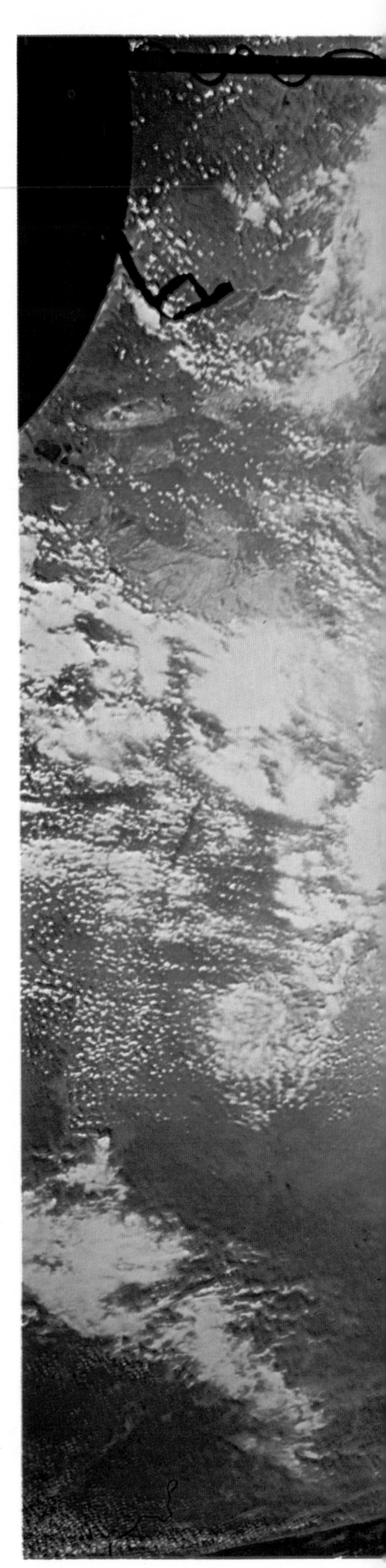

This photograph was taken at an altitude of 620 kilometres (390 miles) from the spacecraft Gemini 11. It shows one of the newest seaways in the world which is at this moment being formed by the drifting apart of the continents of Africa and Arabia. Trace the outlines on the left and on the right and then see how well they fit back together. The Afar Triangle is midway down the section of African coast shown.

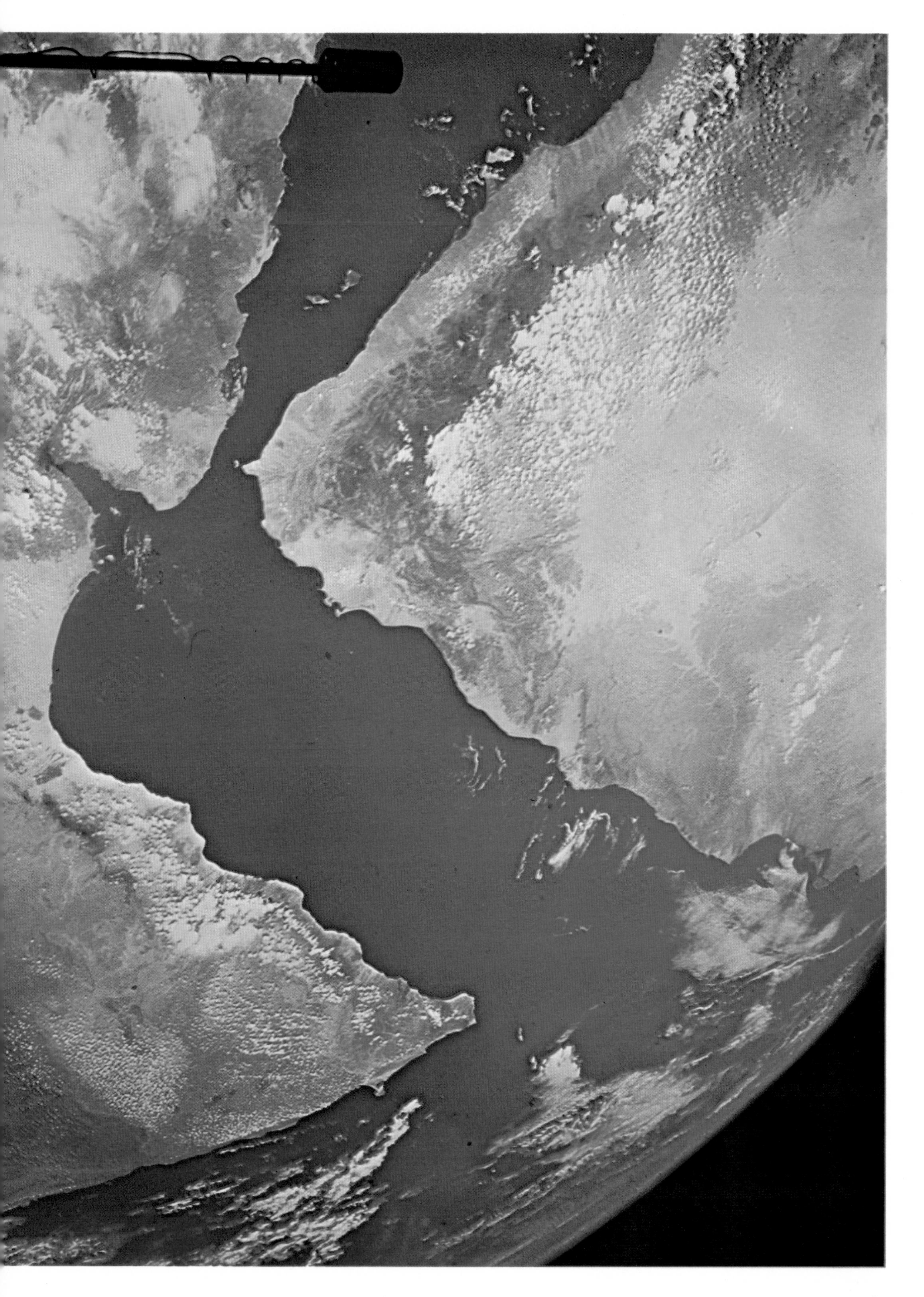

a moonscape of rock pinnacles that barred the road from the lake, each one the remains of a hot mineral spring that had bubbled up into the evaporating sea. Each one a chimney of new rock with internal passageways that led back down to the hot spots beneath, and, to demonstrate it, some were gently fuming into the hot dry air. The eyes of the expedition team became square with success as I climbed to the top of one of them. Eureka, we had found our location, a vent leading down into the bowels of the earth, and from it an issue of sulphurous air, damp with the promise of water vapour. The mix was right, for there around the vent, a tiny oasis watered by volcanic mist, was the unmistakeable blue-green trademark of a life form whose record in the rocks goes back at least 3000 million years.

Wherever there is potential it will be used, and the potential of the extreme environments found on earth is still exploited by the direct descendants of the original organism that had the ability to fix the energy of the sun, using some of it to produce more of itself and releasing free oxygen. They were the fittest then, fit to start the green revolution that would, in time, conquer the earth, and today they are still the fittest in those environments where they have reigned supreme throughout evolutionary time.

Taking great care not to disturb the tiny microcosm of life, I took a handful of rock, crumbled it, and, turning to the emptiness of the desertscape, said 'there are all the potentialities and the limitations of life' and letting it blow from my hands into the hot, dry wind, said 'dust to dust, I am no more than these chemicals borrowed temporarily from the earth and its atmosphere, and compounded by the laws of the universe into the chemistry of life'.

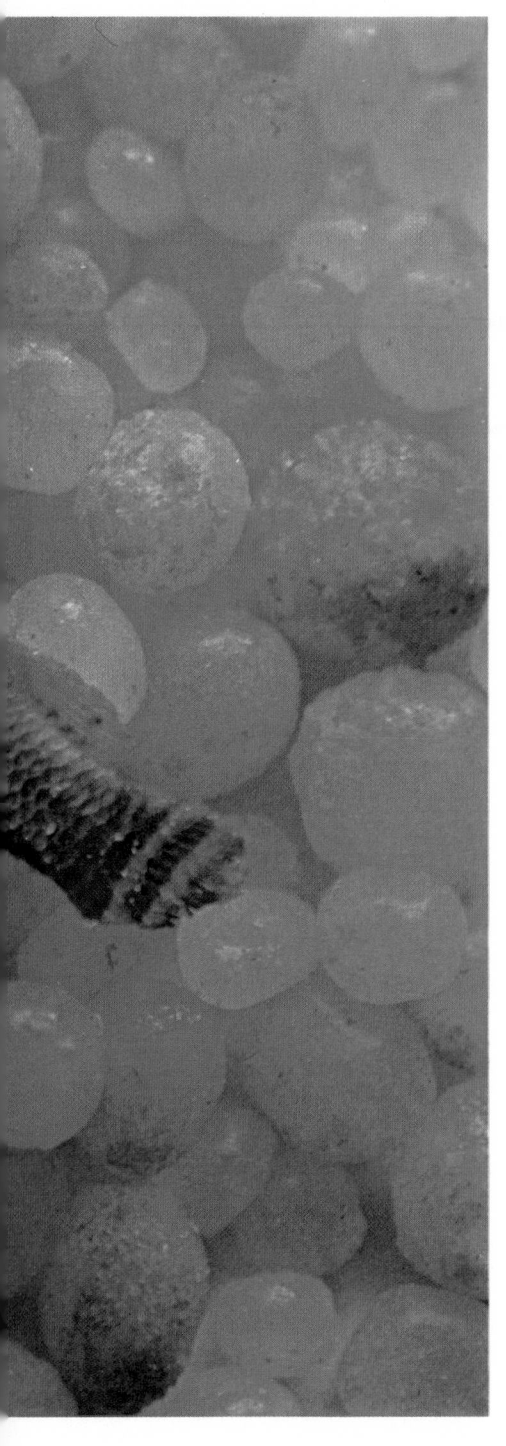

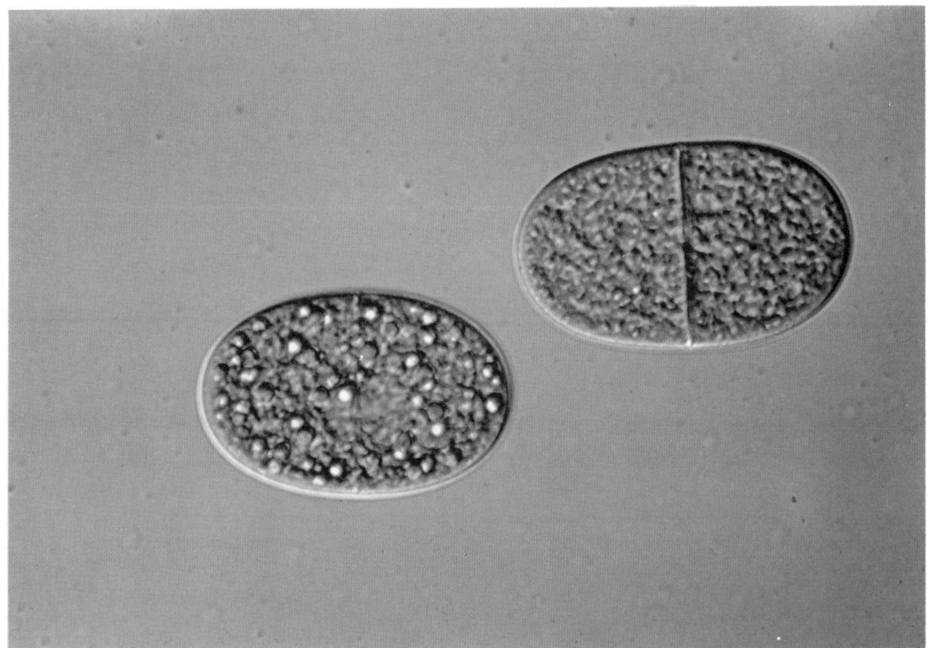

Left Salt marbles and dead fish from the edge of a drying sea.
Above Two cells of a large coccoid blue-green alga (× 1500). One cell is in division and has formed a cross wall (cf prokaryotic cells on p32 which could well be a picture of their ancestors).

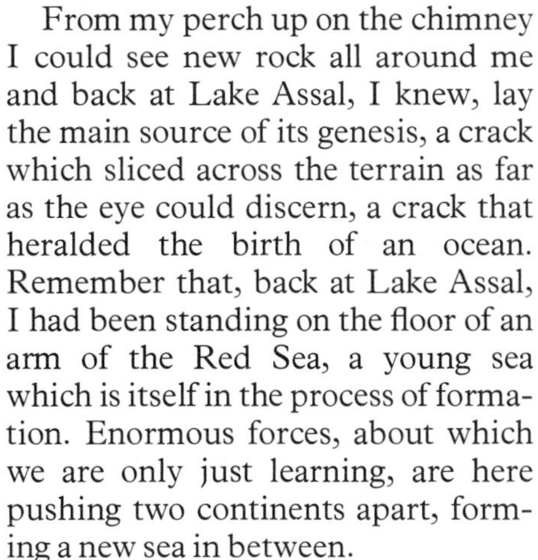

From my perch up on the chimney I could see new rock all around me and back at Lake Assal, I knew, lay the main source of its genesis, a crack which sliced across the terrain as far as the eye could discern, a crack that heralded the birth of an ocean. Remember that, back at Lake Assal, I had been standing on the floor of an arm of the Red Sea, a young sea which is itself in the process of formation. Enormous forces, about which we are only just learning, are here pushing two continents apart, forming a new sea in between.

How do I know? Well, our guides on this particular trip were two French geophysicists whose current task is to measure the rate of the process. Working in the Afar Triangle they have got the best laboratory in the world, for in most other places the mid-oceanic rift is a long way underwater and is, therefore, much more difficult to study. I moved over to the crack and stood with one foot on Africa and the other on Arabia (geologically speaking). Our guides assured me that if I stood absolutely still for a year, my feet would move apart by about 5 centimetres (2 inches), and they should know because at regular intervals they use infra-red lasers to measure it–space-age gadgets for a space-age job. Their work is just part of an all-out effort of worldwide research, which is aimed at a full understanding of the process which moves continents and creates seas.

We now know that evolutionally speaking, not all that long ago there was just one supercontinent, which has been called Pangaea. Around 200 million years ago it began to split asunder and the pieces of the world's largest jigsaw puzzle began to move their ponderous ways to take up the positions we find them in on our contemporary maps.

Proof of their progression is recorded in the rocks, evidence of a very different sort from the fossils which appeared in the rocks of the Cambrian period. The evidence of the movement of Africa and Arabia was in the new rocks under my feet; and the youngest ones were closest to the crack itself, for it is in part the upwelling of new rock that helps to push the continents apart. Shortly after solidification of the rock, while it is still cooling, certain iron-rich minerals present in the new rock will be magnetized, the iron becoming aligned with the earth's north-south polarity. This geomagnetism is from that moment a fixed property of the rock, ideal for use as a direction indicator with which to map out the movements of the continents. Add to this the complicating fact that on at least 100 occasions in the last 70 million years the polarity of the earth's magnetic field has reversed and you will see that mapping the drift of the continents must be a nightmare of orienteering. However, as the continents are eased apart, a geomagnetic record of progress is laid down in the new rock; a magnetic tape which started recording way back in the last days of Pangaea.

We shall see later in the book that, as far as evolution is concerned, this record is top of the rocks, and so it has been for 200 million years, for it has provided more opportunity for the evolution of plants and animals making the earth a fitter place for life.

Continental drift

The biggest jigsaw puzzle in the world

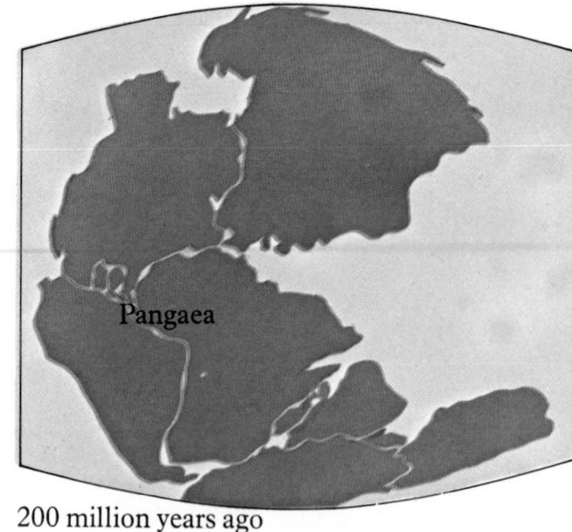

200 million years ago

Cheat sheet for continental drift puzzle

Take a really close look at a map of the world and try to fit the continents together. Some of them, like Africa and South America, are easy, but with the aid of the 'cheat sheet', the rest should not be too difficult.

The evidence for continental drift was, however, not just limited to the shapes of the pieces; the picture on the face of the puzzle also began to fit. Landforms, rock strata and fossils were found in the correct places and the contemporary distribution of some modern animals and plants could best be explained by the assumption that all the continents had once been joined into one great land mass which was called Pangaea. One of the key pieces of evidence came when large deposits of coal were found in the Antarctic and in other regions of the world where it is today too cold for the growth of the coal-forming swamp forests. Either the climate of the earth had changed dramatically over the past 300 million years or the coal-bearing continents had moved, and if this had happened so recently it was argued that they were probably still on the move today.

Continental drift is also known to have occurred before the formation of Pangaea during the Carboniferous period. Alfred Wegener, in 1924, inferred that during the Palaeozoic the continental land masses formed a single continuous unit called Pangaea. Evidence now tells us this was not so and that there existed two super-continents, Laurasia (incorporating North America, Europe and most of Asia) and Gondwanaland (comprising South America, Africa, Antarctica, Australia and peninsular India).

In late Precambrian times the super-continent Gondwanaland was in existence but the components of Laurasia were widely scattered on the earth's surface. During the Palaeozoic the western parts of Laurasia and and Gondwanaland were closely connected but an ancient ocean, Tethys, separated them to the east. Palaeomagnetic evidence also tells us that Siberia and Europe moved independently of one another at this time before being welded together during the formation of the Urals mountain range. By the end of the Palaeozoic era the super-continents were composed of ancient land masses fused together by mountain chains.

Recent studies of the ocean basins and the distribution of active volcanoes and earth-quakes led to a revolution called the theory of plate tectonics which shook the very foundations of the science of geology. There is now no reasonable doubt that the upper crust of the earth is made up of gigantic plates which float on and are gradually being eased apart by convection currents in the molten magma beneath. The resultant cracks, the most spectacular of which run down the middle of the great ocean basins, are being

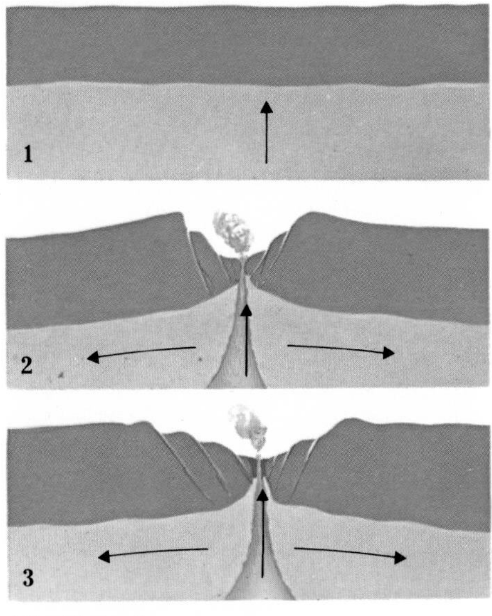

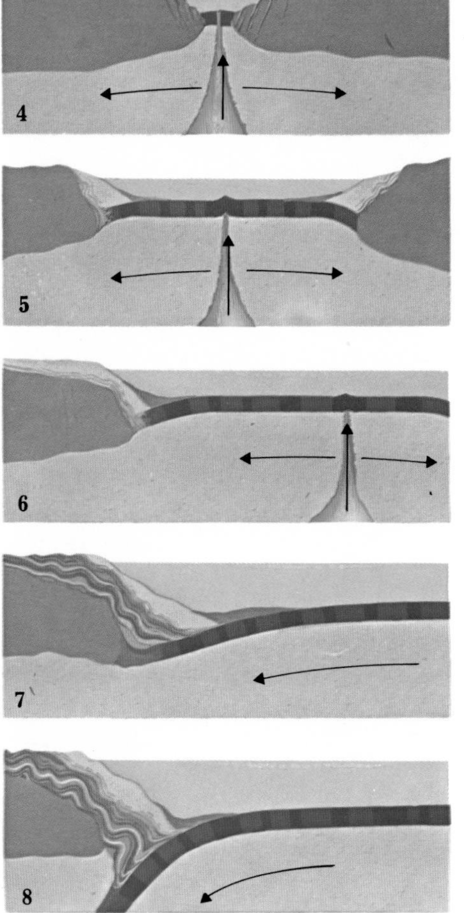

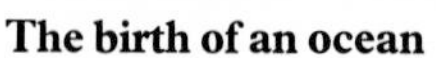

The birth of an ocean

1. Continental crust. **2.** New volcano as cracks form in the continental plate; subsidence and faulting produce a rift valley. **3.** New rock wells up and the two sides drift apart. **4.** The water floods in and an ocean is born; remember the Bible story of Noah and the flood – makebelieve or historical evidence? **5.** The ocean continues to expand and the magnetic tape records its progress in the volcanic rocks of the ocean floor. **6.** Massive sediments begin to accumulate at the margin; sedimentary rocks are formed. **7.** A continental shelf in the making. **8.** New ocean crust dips down below the continental crust; sediments trapped in between may be buckled upwards to form a new range of mountains.

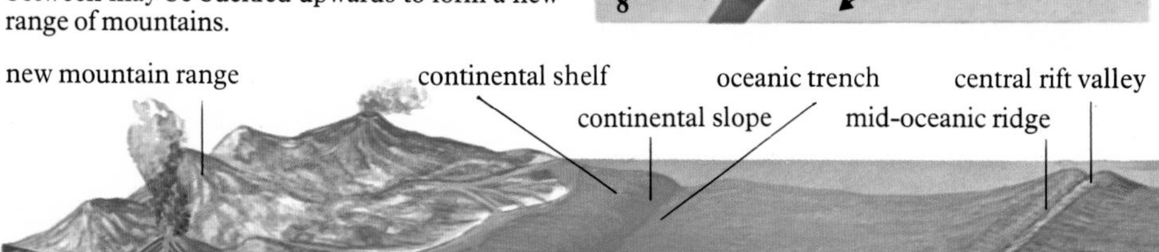

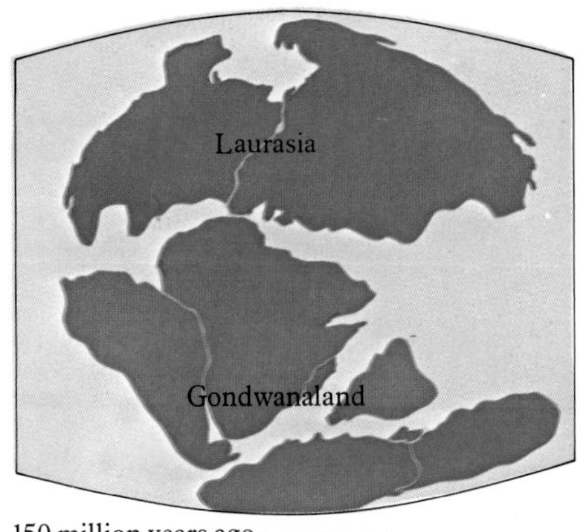

150 million years ago

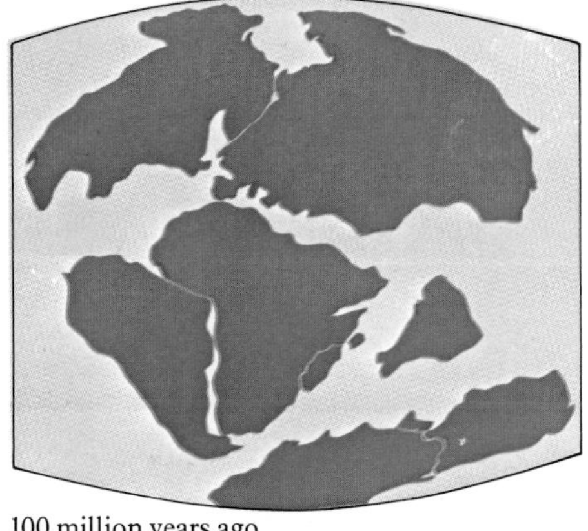
100 million years ago

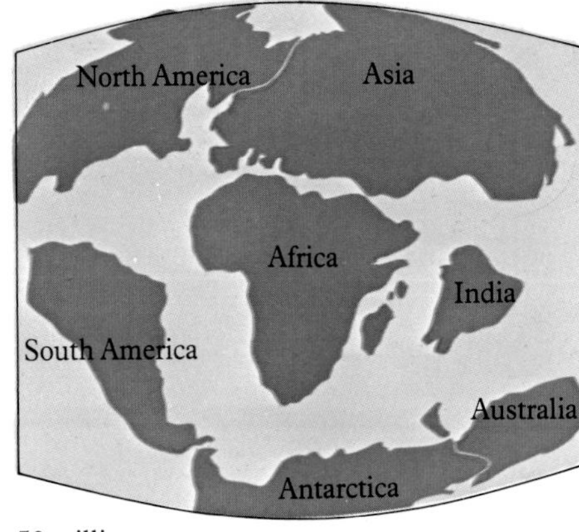

50 million years ago

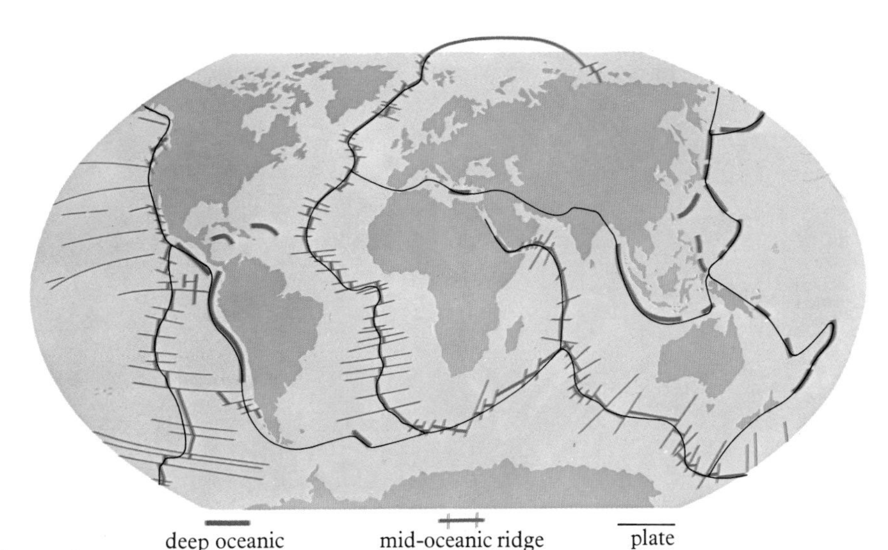

infilled by molten rock welling up from beneath from a series of submarine volcanoes. These form the mid-oceanic ridges which are the greatest mountain ranges in the world.

Where two plates are in collision one usually buckles upwards to form spectacular mountains like the Alps, Himalayas and Andes and the other dips down to produce the massive ocean trenches. Such lines of cataclysmic collision are marked out by being earthquake centres and are often associated with active volcanoes. The most elegant evidence in support of the theory has come from studies of the geomagnetism of the floors of the oceans. As the new rock which pours out from the mid-oceanic ridges starts to solidify, any iron-bearing minerals present in it become magnetized in the direction of the earth's magnetic field, thus producing a wide magnetic stripe. The fact that the earth's magnetic field has reversed at irregular intervals is recorded in a series of such stripes, each successive one magnetized in the opposite direction. These changes may be located from the surface of the ocean as a variation in the strength of the contemporary magnetic field. Knowing the age of each of the reversals, a time scale may be calculated for the process of continental drift. Not only do the shapes of the continents fit but so do all the other facts. Of all the stories of scientific discovery this more than any other had depended on the detailed work of the world-wide community of scientists and its implications are certainly the most earth shattering.

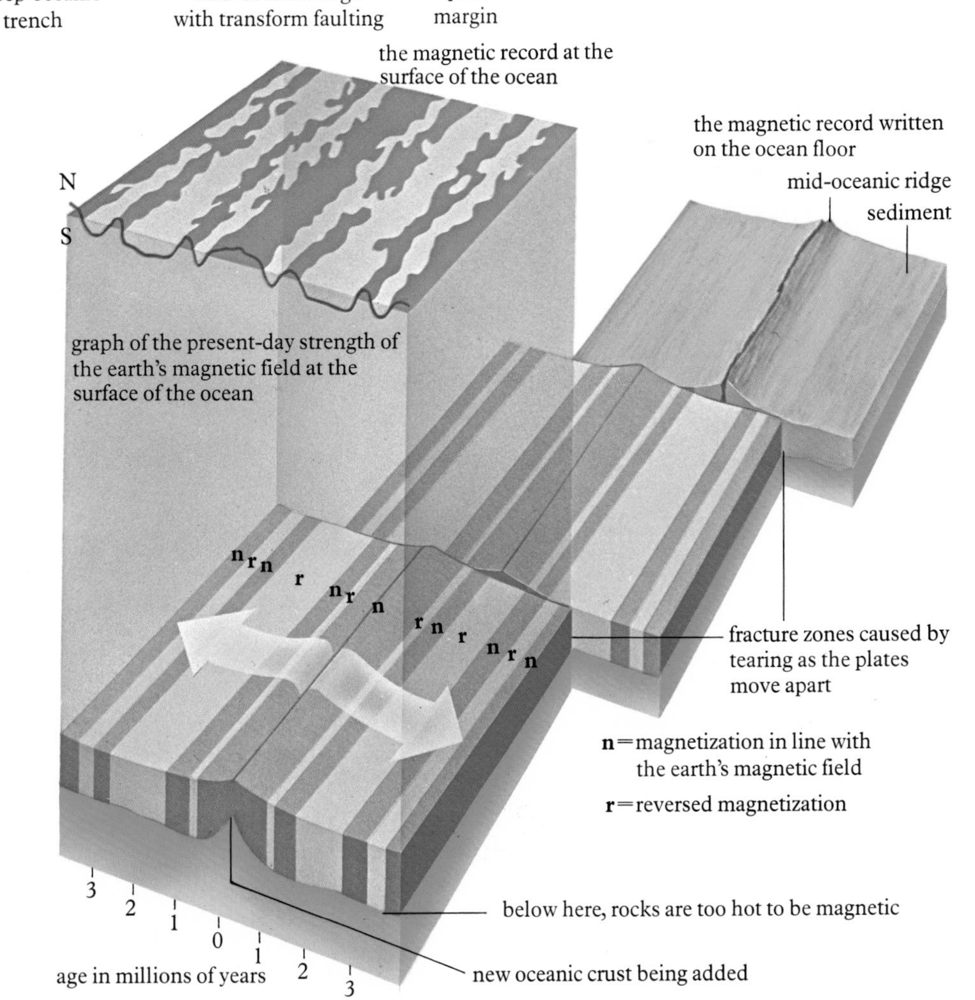

Chapter 3

Living water

Water is the most abundant substance on the face of the earth and, as far as the chemistry of life is concerned, it is the most important for without liquid water there can be no life.

Water is a liquid over the normal temperature range experienced across the bulk of the surface of this planet. That is one reason why the earth was fit for life and life first fitted itself to exploit the watery environments of the sea.

We examine the stages in evolution which filled the sea to overflowing, reaping the full potential of everything it had to offer, and look at the results in the most productive marine system on earth – the Humboldt current off the coast of Peru.

Now you know what you are made of, what about your name? I suppose I am one of the lucky ones because I like mine. However, I know a lot of people, including two of my own family, who feel that they have been a little hard done by.

Every one of us has two sorts of names, the ones that were dreamed up by our parents, and hence are called given names, and the others, which may be single, double- or even triple-barrelled, which we inherited. The latter are our family names and as such they show our links with the past. The fascinating thing is that the further we dig back into the past, the more people we find to add to our list of relations. Perhaps we should use

You may or may not share this crowd's interest in jazz, but you certainly share their ancestry, biological make-up and hence their title, *Homo sapiens*.

all their names. The only problem is to know where to stop, and if you are a biologist that is a real problem.

It was a Swedish naturalist, Carolus Linnaeus, who first formalized the business of name dropping and made all biologists class, well, at least classification, conscious. In Linnean terms your name is Animalia, Chordata, Mammalia, Primates, *Homo sapiens* and the diagram (p 42) shows you some of your relations which include me at the first level– 'Hello, fellow species'.

As there are more than 3 500 million other members of our species alive at this moment, perhaps David James Bellamy is a more useful name; at least it allows me to be picked out from a very large crowd. Unfortunately, it only gives me information about my immediate ancestry and perhaps it would have been to my advantage to know how closely I was related to Paul Getty.

My biological name when reviewed under the successive umbrellas of classification, kingdom, phylum, class, order and genus, tells me a lot about my ancestral line. However, not quite all because, as we have seen in the last chapter, at one time there was only one sort of life. So perhaps we should erect another all-embracing umbrella–a super-kingdom called Life.

What Linnaeus did was to provide biologists with a system of naming, a special language with which we could talk about living things getting their names right every time. It was this clear understanding that helped biologists like Buffon, Lamarck, Wallace and, most famous of all, Charles Darwin to come to conclusions re-

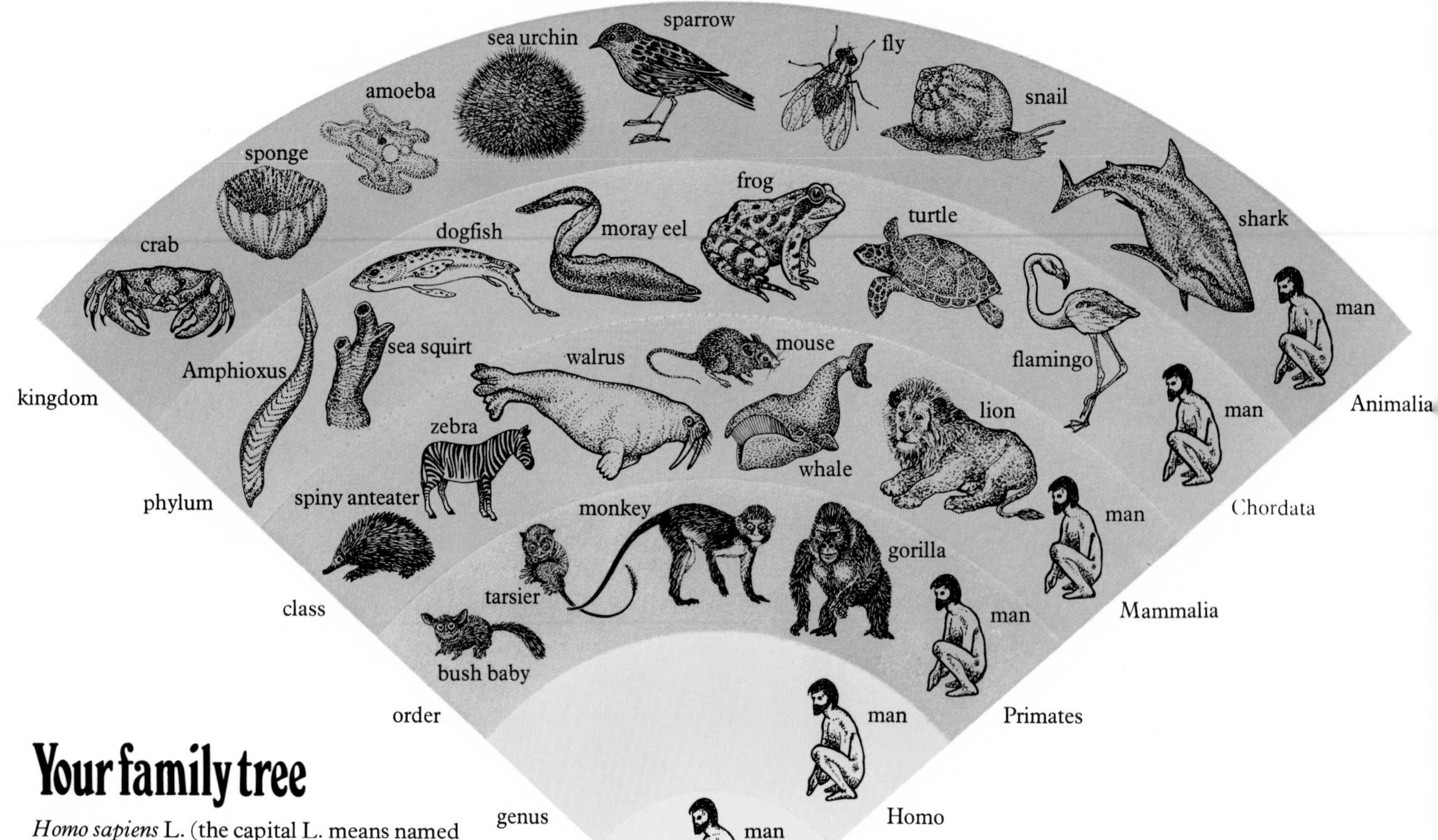

Your family tree

Homo sapiens L. (the capital L. means named by Linnaeus who was one of us too) belongs to the order Primates of the class Mammalia. Despite the many differences which allow us to distinguish between the members of the various umbrellas of classification they all have much in common. They, I mean we, all share the basic biochemistry of life which has been passed down in coded form ever since the evolution of DNA made the conservation of information, and hence the process of evolution, more probable than the chance happenings which could destroy it.

garding their ancestry and lineage. These conclusions led them to formulate the theory of evolution by natural selection.

When Linnaeus first dreamed up his classification, nothing was known about evolution. His reasons for placing different animals in different categories were based purely on their shared similarities. All the animals that are placed in one of the nesting umbrella groups have certain key features in common. *Sapiens* is the species for us, and, as we reckon that we are unique, we have a genus, *Homo*, all of our own. Primates are complex organisms with dextrous fingers and exceptionally well-developed cerebral hemispheres (the 'brainy' parts of the brain). Like all other mammals, primates have hair and the ability to suckle their young with milk. All mammals have a stiff rod to their backs, which acts as a support for their bodies, a feature they share with some pretty strange relatives like the sea-squirts and the lancets. The chordates are animals because they lack certain characteristics that are the trademarks of the plants. For instance, they do not photosynthesize and they actively move about searching for food.

It must, however, be stressed that the absolute distinction between the members of the two great kingdoms of plants and animals is sometimes difficult to make. The reason for this is that back at the dawn of evolution little or no novel changes had occurred and all organisms had much the same set of basic characteristics. Our absolute ancestors were those systems of living chemicals which had the ability to degrade energy and in so doing make more living chemicals.

So we come back to the dead earth from which the raw materials of life are borrowed, and when it comes to names it is here that we have made our biggest mistake. We live on a planet five-sevenths of which is covered with water. Why did we call it earth?

The commonest substance on the face of the earth is a compound called water and it is a very important compound, especially to us living chemicals. If you could take the two most reactive substances on earth, dry them off completely and then mix them in the complete absence of water nothing would happen. In the absence of water, chemical reactions can not take place, or perhaps it is more correct to say that they would take place so slowly that their progress would be undetectable. Water is, therefore, looked upon as the universal catalyst, which is a substance which speeds a chemical reaction without itself being changed.

No wonder then that water and life are inseparable. In fact, a living organism can be regarded as a mass of water within which life chemicals are structured for efficiently flowing energy and producing more of themselves. In a nut shell, the catalyst with the self-fattening centre. If you have a weight problem, next time you look despairingly in the mirror just imagine what you would look like minus all your water. From a not so trim 100 kilograms (220 pounds) you would desiccate to around 20 kilograms (44 pounds) in weight, a neat pile of dead, dead dust. The same goes for almost every other living thing, showing that water is the most abundant constituent of life. Perhaps it is not so surprising then that evolution first got cracking in the sea, and to prove it the vast majority of the world's first abundant fossils are of marine animals and the rest of calcareous seaweeds.

The fauna of the Lower Cambrian period, those 30 million years when it all started to happen, consists of at least 500 species being representatives of no less than eight of the major phyla–Protista (single-celled animals), Porifera (sponges), Cnidaria (jellyfish, sea anemones, sea pens and corals), Mollusca (snails, mussels and squids), Annelida (segmented worms), Brachiopoda (lamp shells), Arthropoda (crabs, lobsters and barnacles) and the Echinodermata (sea-urchins and starfishes). The examples given in brackets after the proper names are examples taken from the contemporary fauna of the oceans so you can see that at the phylum level nothing has been lost. The only major animal group that is missing from the early Cambrian fauna is the Chordata, members of which are only found in the fossil record much later. It would thus appear that the bulk of the main steps of animal evolution was complete almost before it had a chance to begin.

Why this sudden burst of diversity? The potential was there and it had to happen. At least five-sevenths of the energy that falls on the earth falls on the surface of the sea, which contains water and dissolved minerals in plenty. The sea was devoid of life, there were no competitors, and so the diverse environments of the oceans were up for grabs. The race was on and it was a race in which everyone could be a winner, each new organism adapted to do a particular job in the evolving society of the sea.

Once there was chlorophyll the energy of the sun could be tapped, once free oxygen was being continuously produced that energy could be used more efficiently, once there were plants there could be animals to eat them and the plant-eaters provided opportunity for the flesh-eaters to evolve; once upon a time it all began to happen.

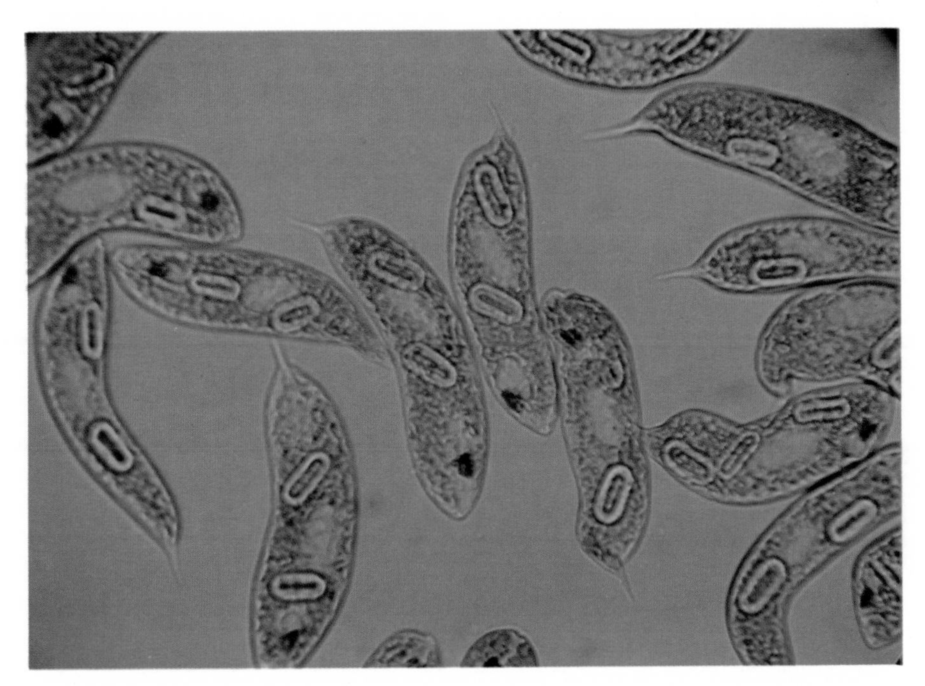

This group of eukaryotic cells are green protists called *Euglena*. They are of common occurrence, especially in farm ponds and puddles which are enriched by organic waste. Although a green plant, *Euglena* has the capability of actively swimming through the water by means of a long whiplash flagellum, the bases of which can be seen in the picture. Inside each cell are many substructures called organelles. For example, the small green discoid structures are chloroplasts which fix the energy of the sun; the large doughnut-shaped objects are starch stores; and the orange-red areas are light-sensitive eye spots which help to guide the 'animal' along.

Is it a plant or an animal? When living under certain conditions it can forsake the process of photosynthesis and feed by ingesting organic particles into the cell, so since it can not make up its own mind, why should we worry?

The early Palaeozoic sea

I would have liked to have been beside the seaside then. I wonder what a giant nautiloid was really like, and what groups of animals lived and yet left no imprint in the fossil record?

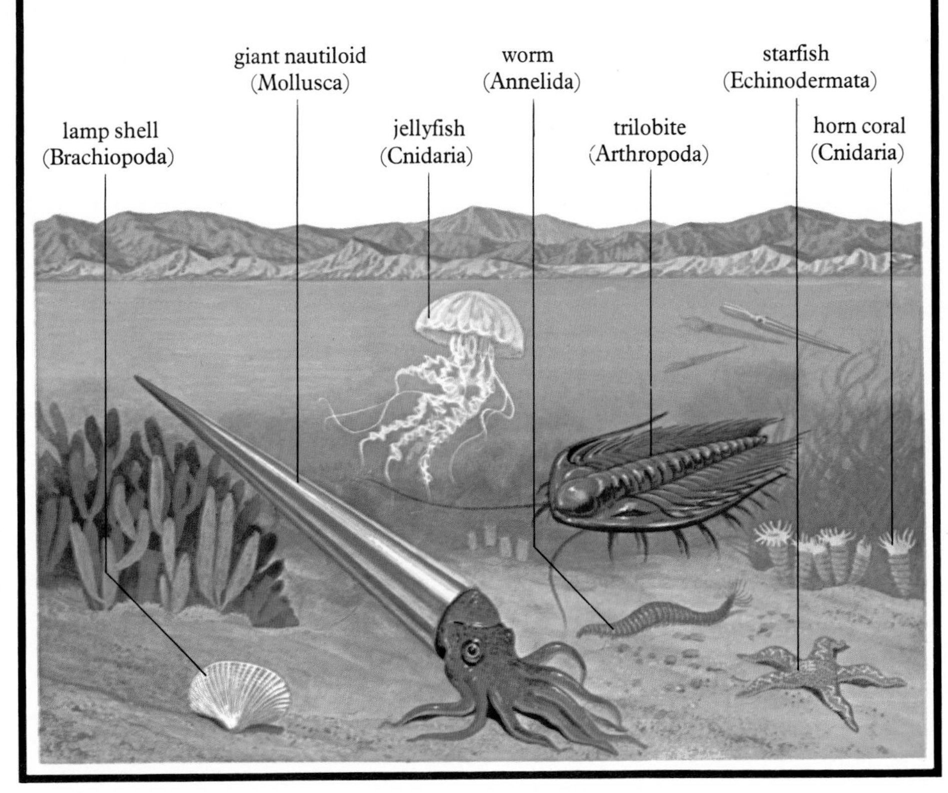

The Protista

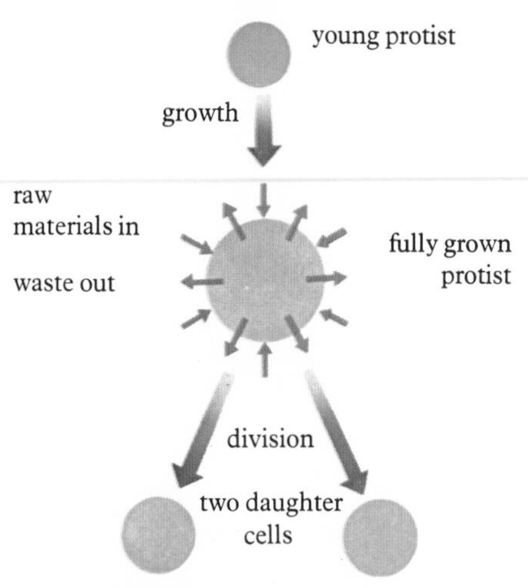

The majority of the protists are too small to be seen with the naked eye. Their limited size is due to the problem of packing all the life functions into a single cell. The larger the cell, the more raw materials it will require and the more waste it will produce. However, the living volume of the cell increases more rapidly than its surface area through which all exchanges must take place, hence the limitations of size. There are many ways in which life has got over the limitations of the scale effect but as far as the protists are concerned the simplest way is to keep a trim healthy figure by dividing to restore the norm.

Success at each stage was dependent upon getting the design right. It would be useless being a browser if you could not move about to browse, or a predator if you could not outrun or outwit your prey. The adaptations which allowed an animal to get over its problems conferred fitness on the individual and its descendants, and provided potential for the future.

The major breakthroughs are recorded in the characteristics of each phylum, so here is a potted version of the first major phase of evolution.

The Protista are complete animals in a single cell, or to look at it another way single cells which are capable of carrying out all the basic life functions. These are:

(a) *respiration*, the mechanism by which the energy stored in food is made available; (b) *growth*, the use of part of that energy to increase their mass; (c) *excretion*, the mechanism by which toxic wastes are voided from their body; (d) *irritability*, the ability to respond to stimuli from the outside environment; (e) *reproduction*, the channelling of some of the energy into the production of more animals of the same sort.

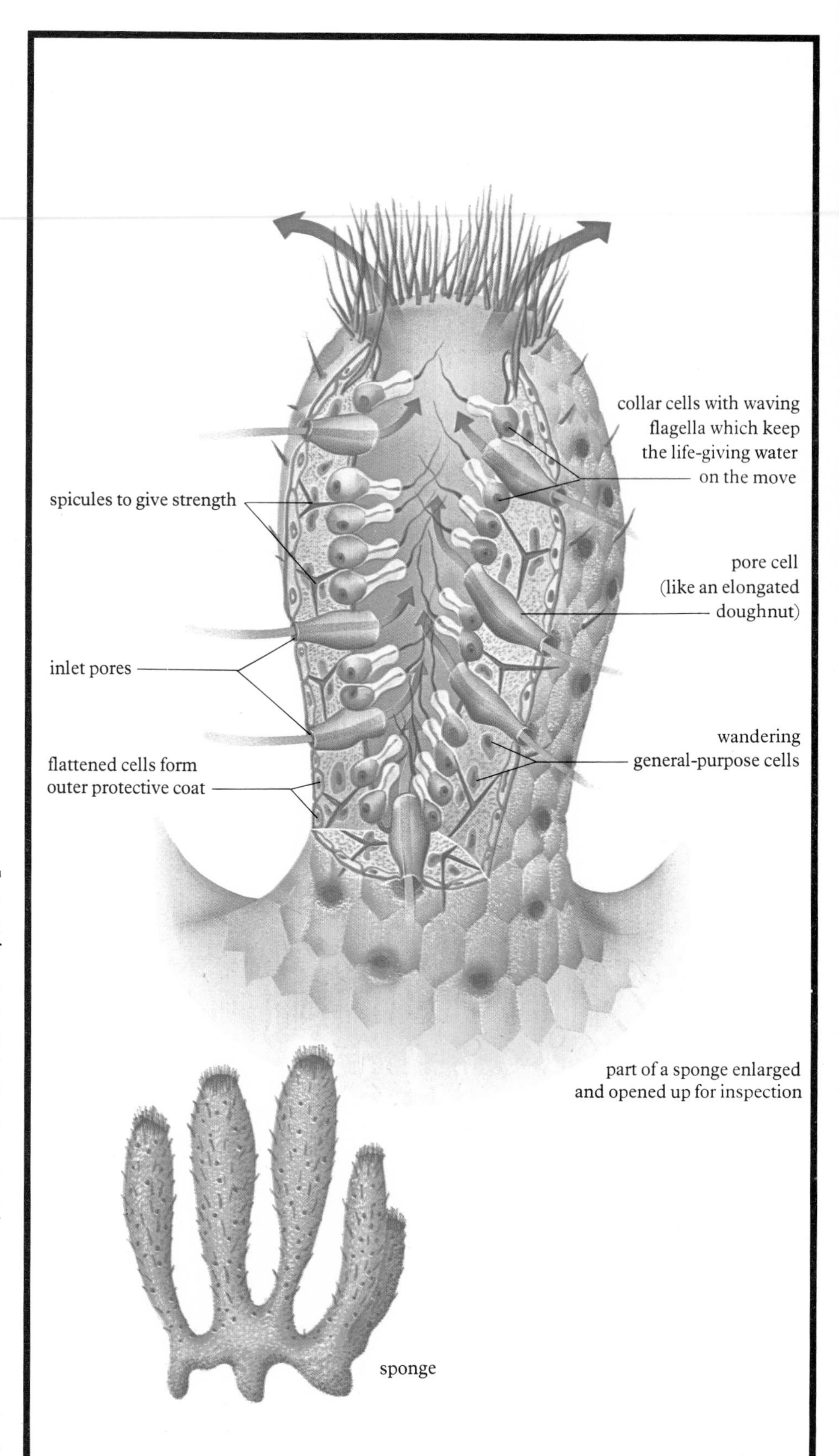

part of a sponge enlarged and opened up for inspection

The Porifera

The sponges are looked upon as the simplest of the many-celled animals (Metazoa). They have overcome the problems of the scale effect of the protists by being no more than a bag of holes through which the environment (the water in which they live) is continuously flowing, thus providing all the component cells with an adequate supply of food and a way of disposing of their waste. Although they have lived in the waters of the earth since the dawn of evolution they are still abundant and important components of life in the modern seas.

The medusa of *Gonionemus*, a free-swimming member of the Cnidaria.

I know you and I can do a lot more than that but if those were your sole accomplishments you could still be said to be alive. The all-purpose cells of the Protista can do just that.

From here on up all the animals consist of many cells, some of which are highly specialized and all of which work together. They are therefore known as Metazoa.

The Porifera are the sponges and each one is a mass of many thousands of cells, some of which are specialized to do a particular job. They are all capable of all the general purpose 'core functions' of life, but some are modified to perform one function extra well.

The Cnidaria show a further stage of specialization in which certain types of cell are grouped together into working units called *tissues*. For example, the cells that form the outer layer of their body are modified for protection, forming a tissue called ectoderm. In contrast, the cells that form the inner layer of their sac-like bodies are modified for feeding and form a tissue called endoderm. With such a high degree of specialization there is a certain amount of limitation in their 'core' life functions, so much so that here at the tissue grade of construction we find special cells set aside for reproduction.

From here on up all animals may be said to have reached the *organ* grade of construction because in all we find specialized cells working together in specialized tissues, a number of which work together to perform a specific function or functions.

This high degree of work-sharing was only obtained after evolution had produced a new type of cell–one that would never have direct contact with the outside environment or for that matter any of its products–not ectoderm nor endoderm but a new layer, the mesoderm. It was evidently the mesoderm that put the filling in the sandwich of life for it is mesoderm that helps to form many important organs.

Such a radical change must have both advantages and disadvantages. First the good news–the new type of cell will always be shielded from the vagaries of the outside environment and will have a new set of more stable conditions, a new environment in which to be irritable. Now the bad–the cells will not be able to fend completely for themselves in feeding and excretion and will be dependent on other parts of the system. There will be a need for organs to sense the outside world, a system of nerves to pass on the information, a blood system to transport food and waste around the complex body, and kidneys to deal with the waste, thus ridding the corporate body of the toxins produced by the life of each cell. All the component cells, including the mesoderm, have the ability to carry out most of the 'core' life functions but with specialization comes complexity, and with complexity comes dependence on other parts of the system. (Just think how nice it is to have a washing machine–well, at least until it goes wrong.)

What then was the advantage that the organisms gained with their third layer of cells? In a nutshell it allowed them to grow larger and move faster. The maximum size for a living protist is a function of the volume of its cell, its pace of life and the surface area through which raw material must be imported and waste ex-

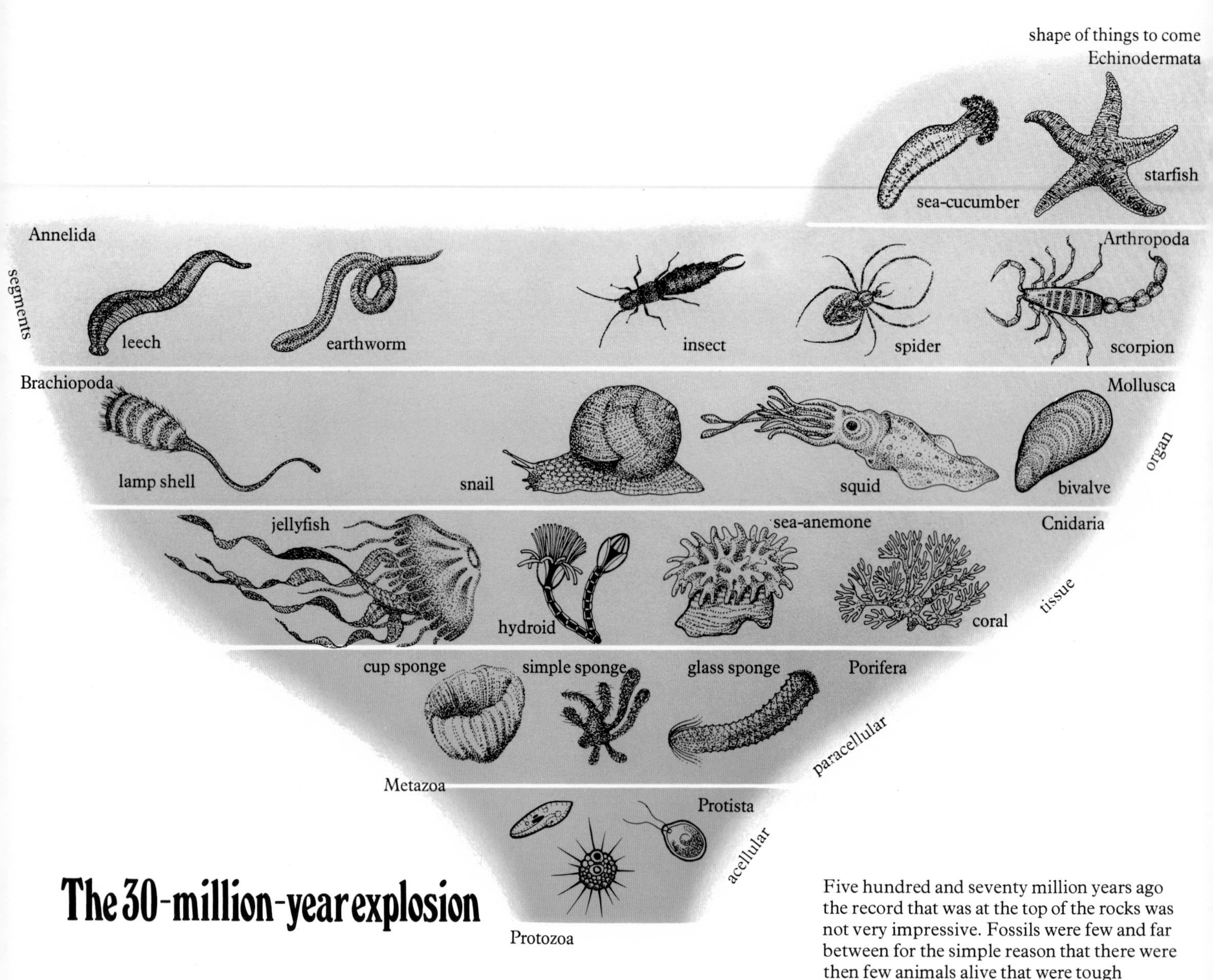

The 30-million-year explosion

Five hundred and seventy million years ago the record that was at the top of the rocks was not very impressive. Fossils were few and far between for the simple reason that there were then few animals alive that were tough enough to be fossilized. Thirty million years later it was a very different score – eight of the contemporary top ten animal groups were solidly in the charts.

ported. If the ratio is overstepped the pace of life must slow or the organism will die, unless it divides to reinstate the balance.

The sponges can cope with their great mass of cells because most of them are kept in direct contact with the sea water which continuously flows through their myriad pores. The Cnidaria, with the basic, two-layered structure, were stuck and their individuals are either small or, if large like some jellyfish, their bulk is made up of a non-living packing material. Being non-living the jelly does not require much in the way of services but it can not contribute much to the life of the organism. This is one reason why jellyfish are such poor swimmers and so live at the mercy of the environment which often leaves them on the beach high and drying, and for a marine creature that means dying.

The Brachiopoda. Although, today, the lamp shells are both small and unimportant members of the marine fauna, they were once much more abundant and some were much larger. They lead a sedentary life, safe within their protective shells, and yet they have the makings of all the organs and systems that are found in more advanced groups. They also have a body cavity or coelom. Some of their bulk is composed of a fluid-filled cavity which was formed as a split in that all-purpose layer, the mesoderm. Like the packing tissue of the jellyfish, the fluid adds bulk without requiring much in the way of food or removal of waste. It also acts as a hydrostatic skeleton against which the muscles of the body wall can work more efficiently, a property that is very important to some of the more advanced phyla.

Please do not misinterpret this. I am not saying that the protists gave rise to the sponges from which the cnidarians evolved, and so on. All I am pointing out is that each phylum does exhibit stages in the process of evolutionary advance. It is best to look upon each phylum as a highly successful offshoot from the main stream of advance. It is just as if once a major breakthrough had been made it opened up whole new vistas of potential which were immediately seized upon. One of the next successful offshoots is the Mollusca.

The Mollusca range in size from

very small up to the giant squids which may be more than 20 metres (66 feet) from the top of their head to the tips of their tentacles. They have body cavities in which a complex gut is suspended, an efficient heart and circulatory system, advanced kidneys and, among other advantages, a complex nervous system to keep it all under control.

The molluscs are a fantastic design job, custom-built to play a number of roles. Like the lamp shells, most two-shelled molluscs sit and strain their food off from a stream of water that is pumped through their shells. The snails and their like glide slowly about grazing on the plants that grow in their path. In contrast, the squids and the octopuses are efficient predators and it is in this group that we find the greatest development of the nervous system. Brain power exactly where it is needed most, in an animal which must successfully stalk its prey or starve to death.

It is their own special adaptive trademark, a tough protective shell into which they can retreat when danger looms, that marks them out in the fossil record. It is of great interest that in the more mobile squids and octopuses, the shell is either lost or becomes an internal, skeleton-like structure, yet in the Cambrian the squids had massive external shells.

The Annelida missed out on the shells, taking up the flag of the hydrostatic skeleton and wiggling it to their great advantage. Their trademark is segments and their segmented bodies are ideal for all worm-like activities like creeping around and through things. Each segment forms a separate, pressure-tight compartment, a coelomic capsule through which a central gut, blood and nervous system can run, and against which the muscles of the body wall can work, making for efficient wriggling. It seems likely that the worms originally evolved to exploit the potential of the soft sediments that were forming on the bottom of the sea, a job which they still do to slim-line perfection.

The sandworm *Nereis diversicolor* is a segmented coelomate which is beloved of carnivorous fish and hence fishermen. Note the segments, each of which is adorned with two paddle-like appendages.

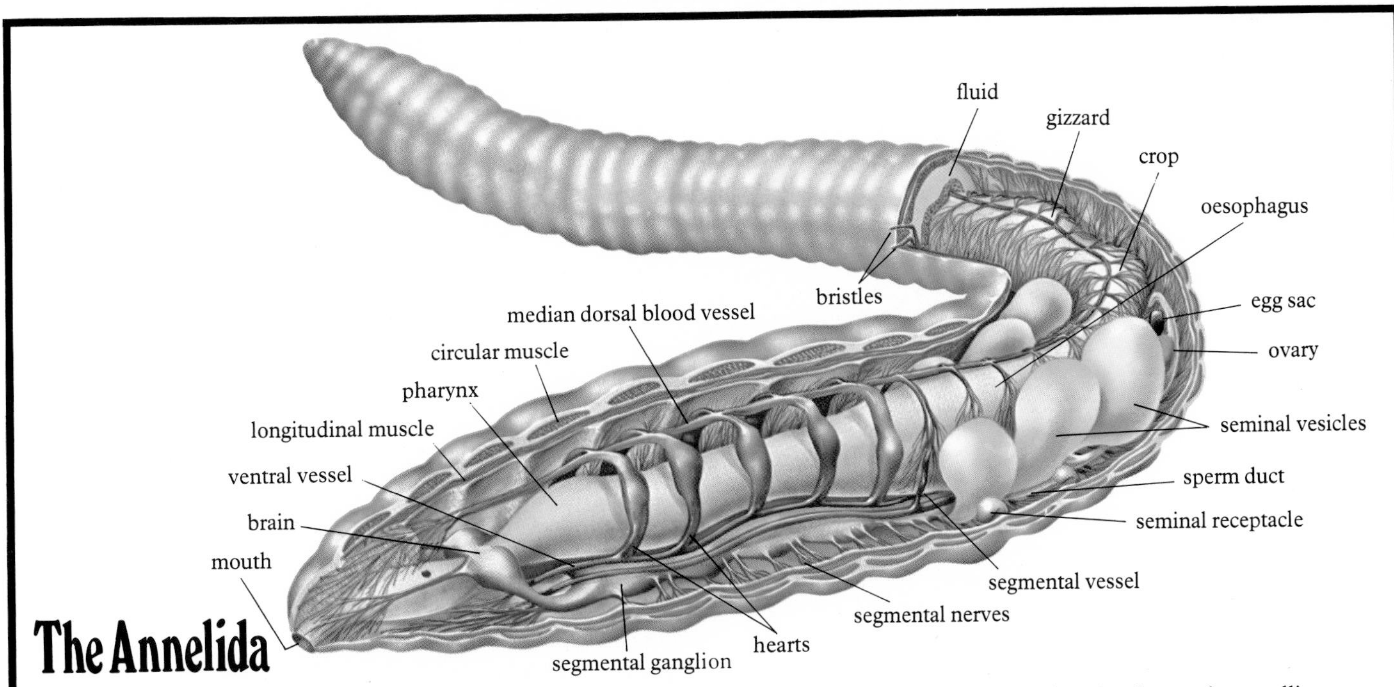

The Annelida

I have a very soft spot for worms, especially the ones that live in my garden, because I know that they are there hard at work tunnelling away in search of food and in so doing keeping the soil well stirred, well drained and well aerated. As a budding biologist you are often given an earthworm to dissect and I must confess that I hated doing it and that is one reason why I became a botanist. This painting captures the importance of the fluid-filled coelom and all the adaptations, innovations and systems which took the evolution of animals another step on the road to success.

Locusts are land-dwelling members of the Arthropoda. If success were measured by numbers and the damage done, then the desert locusts would be at the top of the league. Remember the desert locust is just one of perhaps one million different species of insect which share the earth with the other products of evolution. A plague of locusts is simply a local population exploding in search of more food. Many perish but the few that win through go on to start it off all over again.

The Arthropoda are, on the whole, adapted to a much more active life, living in a variety of environments and moving over a variety of terrains. A segmented body plan is still in evidence, especially in the arrangement of the multitude of legs or rather appendages, for they are adapted to do a whole variety of jobs, from attack to defence, environmental monitoring, feeding, locomotion and communication to baby minding! The success of the group, however, resides on their backs in the form of a highly protective, all-enveloping external skeleton. Throughout evolution, and especially in modern-day terms, they must be regarded as the greatest success story, for not only does the phylum include the Crustacea (crabs, lobsters and shrimps) and the Arachnida (the spiders and mites), but also the class of the insects whose diversity far outnumbers that of all the other animal groups put together.

It would be an insult to call the arthropods an offshoot from the mainstream of evolution. They are one of the terminal groups, the apex, the tops, at least as far as one mainline of evolution is concerned.

We humans share many of the mainstream characteristics of the arthropods–many cells, tissues, organs and a body cavity–but that is all. We do not have any of their specialities which have gained for them worldwide success; we have our own.

We are members of the only major phylum that was missing from the seas of the early Cambrian, the Chordata. Perhaps more work will reveal the presence of primitive chordate forms for they must have been there. Why can I be so dogmatic? Well, we know that the chordate stock split off from that of the arthropods way back at the dawn of the mesoderm. How do we know? The mesoderm of all the phyla that are offshoots from the arthropod line is formed by new cells budding off from the endoderm and the coelom is formed by a brand new split in that mass of cells. In contrast, the mesoderm of the members of the chordate line is formed by budding off from the primitive gut, and as the bud contains a piece of the gut cavity it has a ready made coelom.

It is rather interesting to speculate on the fact that the gut cavity of any animal is not really inside the body of the animal at all. It is simply an extension of the outside world, a hole through a complex doughnut. Food passes through the gut and is digested *en route* before some of it is absorbed into the body, the rest passing out at the other end. This means that the primitive coelom of the chordate line has its roots back in the outside world. Whether any real importance can be attached to the fact that when it comes to the origin of the coelom, one is an outside and the other an inside job, I do not know. What I do know is that it was a main point of divergence in evolution and that is important. If the chordates are missing from the fossil record of the Cambrian, there is certainly one group of the same coelomic lineage that is very well represented.

From a single cell to coelom, an important split in the ancestral stock

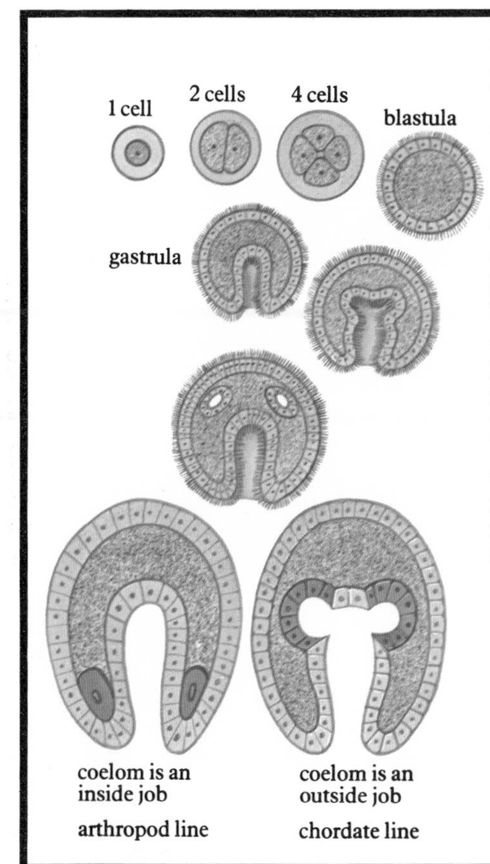

One cell divides to form two, four, eight, a hollow ball of cells (the blastula), and then invagination produces the gastrula which puts a second layer into the shape of things. Up to this stage in our development most of us coelomates look much the same. The most important split comes with the formation of the third layer, the mesoderm, and so it did way back in evolution.

Overleaf The tube feet of a seastar, each one containing an extension of the coelom. The feet are used for locomotion and for opening shellfish, which form the staple diet of the seastar.

The Echinodermata are there in force and to prove it they have left behind tough evidence of their many successes. Their trade marks are armour plates each made of a single crystal of calcium carbonate; a radial symmetry with five sectors arranged around a central part; mobile spines, each set on a ball-and-socket joint; and, in the best-preserved fossil beds, even the frail imprint of an extension of their coelom may be found. Like all the extant forms, such as the sea-urchins, sea-stars, brittle-stars, sea-cucumbers and sea-lilies, those of the Cambrian held on with or moved about on thousands of delicate but highly effective tube feet. The myriad of tube feet are worked by a complex water vascular system which is part of their coelom, a real outside job.

We cannot, of course, discern the method of formation of the coelom by simply looking at adult animals, however well preserved. The only way is to follow the stages of development from the fertilized egg, and it is well to remember that any animal, however complex, starts life as a single cell. The complexities of cells,

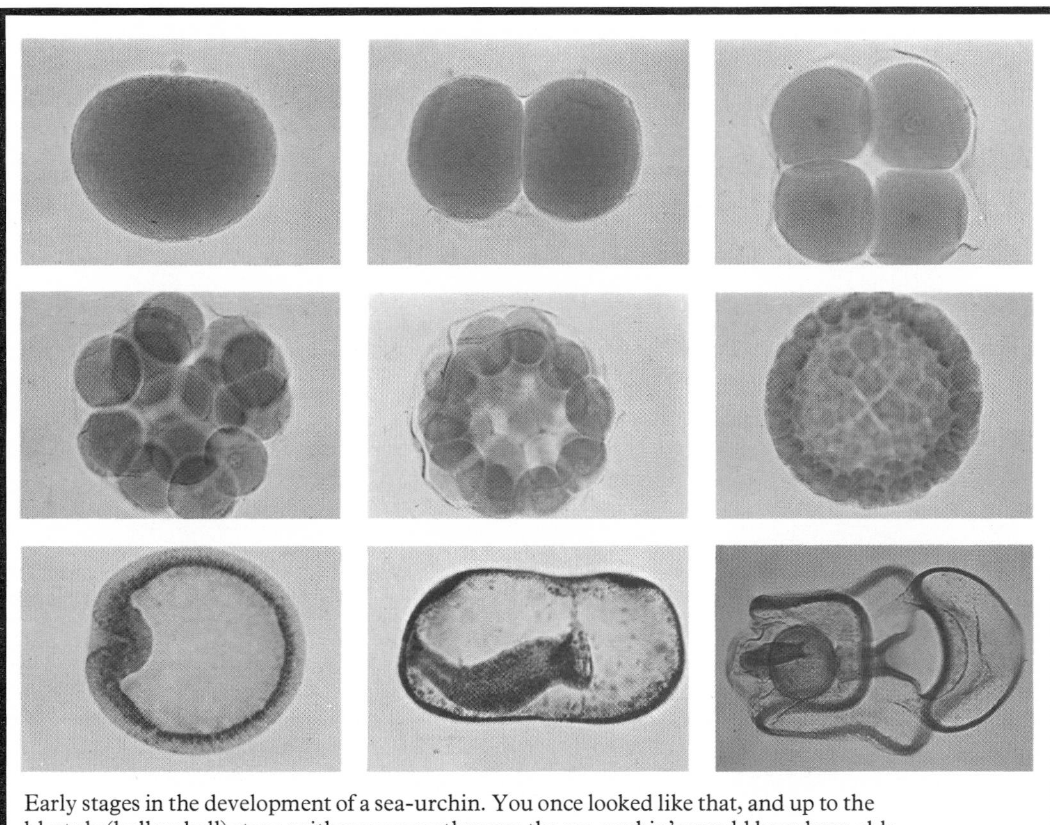
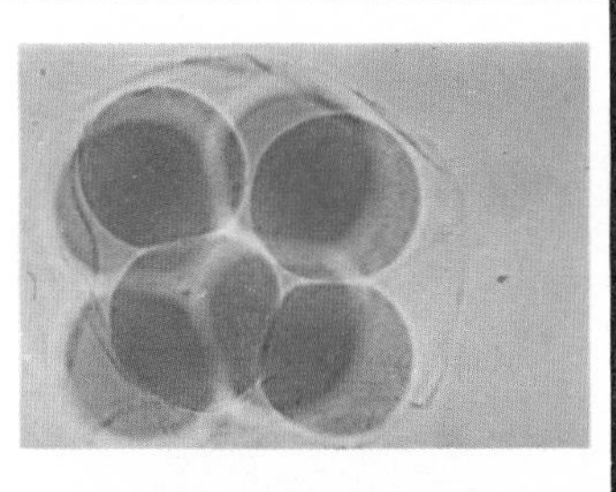
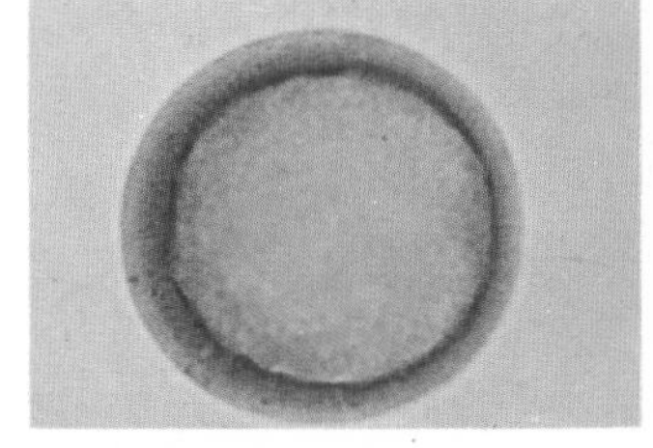

Early stages in the development of a sea-urchin. You once looked like that, and up to the blastula (hollow ball) stage neither your mother nor the sea-urchin's would have been able to tell their offspring apart. The real miracle is that the original cell contained all the information necessary to make the adult urchin. At each division a complete replica of the information is passed on so it should be possible to produce a new individual from any one of those cells. Just what switches on and off certain parts of the information, allowing differentiation and development to take place, is now being actively researched.

1 Fertilized egg with polar body formation
2 The egg divides to form two blastomeres
3, 4, 5, 6 More divisions produce four, eight, sixteen and then thirty-two blastomeres
7 Non-motile blastula
8 Late ciliated blastula
9 Gastrula showing the beginning of invagination
10 Late gastrula, lateral view
11 Ventral view of larva with well-formed bipinnaria

The silica skeletons of diatoms, each of which is made of two halves which fit together like an old-fashioned pill box, have long fascinated microscopists. These have been arranged on a slide with loving care by some master of the microscopist's art. They show some of the enormous range of structure and detailed beauty of this important group of plant plankton (× 450).

tissues, organs, coelom, and so on, are built in during the early stages of its development. It is almost as if the embryology of every animal retraces the main steps of its evolution. Up to the coelom stage, you and I were very like the embryo of a starfish. (Next time you catch one, put it back in the water.)

So, by the end of the early Cambrian the main developments had taken place, the potential was there, the conditions were right, it had to happen and we have the fossils to prove it. The problem is that all they do is prove that it happened; they do not provide the reason why.

If we want to see for ourselves how the various products of evolution are fitted to make the most of the environments on offer, then the only way is to study a contemporary situation, and what better than the most productive marine system.

When Alexander von Humboldt, explorer, naturalist and geographer, first visited the coast of Peru in 1803, he was puzzled by the fact that, though shrouded in mist, the lowlands were a virtual desert. He found the answer in the temperature of the sea water, which by tropical standards is very cold. The reason is that a strong current sweeps up along the coast of South America bringing with it cold Antarctic waters. Onshore winds are, therefore, cooled by the waters of the current, which was later named after Humboldt, and they only begin to warm once they are over the coastal strip. As warm air can hold a lot more water vapour than cold air, this change in temperature virtually sucks water from the soil. Hence the mist and the desert, the water coming down as rain simply going back up as mist.

If the coast is a desert, the same can not be said of the offshore waters for they are teeming with life, and if life did evolve on the edge of the sea it could not have chosen a more spectacular or potentially productive coast than that which is lapped by the cold Humboldt current. Of course back in those far off days South America probably had no free existence; it was part of the land masses which later formed the great continent Pangaea. All the land at that time must have been a total desert for the simple reason that there were then no land plants or animals.

South America is today moving westwards, riding on the back of nazca continental plate, the only trouble being that the floor of the east Pacific is formed by another plate which is moving east. Where they meet you can guess that something pretty spectacular must be happening and you can not get anything much more spectacular than the Andes which rise to a height of 6850 metres (22834 feet) and the Peruvian Trench that dips to an impressive 6000 metres (20000 feet) below sea level. One plate has buckled upwards and, as if to let the steam off, has produced the world's highest volcanoes. The other has dipped

A good haul of anchovies, on the coast of Peru.

under the grinding impact producing spectacular earthquakes and stirrings in the deeps. Spectacular it may be, but why the promise of productivity?

If you are an organism, plant or animal, and you live in the open ocean, your life is one continual problem of keeping up with the plankton, for that is the name given to all the free floaters. For an animal it is not too bad as long as you can live under pressure and in cold water, but for a green plant a sinking sensation can mean certain death because the deeper they go the less light there is for photosynthesis. It is for this reason that many of the plant plankton bristle with spines and curved 'horns' which increase their surface area and hence their ability to float.

It is easy to see how advantage can build on advantage bringing about change. An individual which has slightly bigger 'horns' than the rest will keep more easily up in the swim. It will thus be at a considerable advantage energywise, a fact that could allow more energy to be channelled into reproduction. That means more big 'horns' keeping up in the swim, and so on, generation after generation, a population explosion of the new type of fitter cell. This is in essence the basis of natural selection which Charles Darwin and Alfred Russel Wallace proposed as the mechanism of evolution in 1856. The fact that under normal circumstances we always find a diversity of plant plankton shows that even in the fully mixed media of ocean water there must be a whole host of environments to fill.

In the same way, under normal circumstances the plant plankton feed a variety of animal plankton herbivores, carnivores, scavengers and freefloating omnivores, each playing their own role, increasing the number of pathways through which energy can flow. At each step in the food web there is a loss of potential energy, so much so that the higher you live up the chain the less food comes your way. This is the law of the sea and it warns of diminishing returns, especially for those at the top.

The phytoplankton of the Humboldt current have one special characteristic; most of them are large, colonial forms. They are so large that they can be seen and hence selectively eaten by a local member of the chordate line, *Engraulis* by name, the top of the pizza anchovy. Here we have a fish that is, at least in part, just two steps along the energy chain from the sun. To prove it the local fishery has been estimated in a good year to produce a staggering 35 tonnes of high grade protein per square kilometre (90 tons per square mile). If a beef farmer could get anywhere near that figure he would certainly be king of the 'big horns'.

Long before man came on the scene, another offshoot of the chordate line cashed in on the productive anchovies and made a real mess in the process. Dotted along the coast of Peru are a number of small islands which must rank among the smelliest and noisiest islands in the world, and each one is freshly whitewashed by its inhabitants. The inhabitants are tens of millions of seabirds, the noise is their amorous and aggressive billing and cooing, and the whitewash

is their corporate body of excreta. Birds have evolved a novel mechanism of defecation and excretion in which much of the water is extracted from the waste before it is voided from the body. The resultant semi-solid mass is rich in, among other things, phosphate and nitrates, and is readily soluble making it an ideal fertilizer. The reason that it stays put on the islands is the lack of rain along this part of the South American coast; although it would be more correct to say stayed put, until the dawn of the industrial revolution. This revolution marked a new explosive phase in the evolution of man, fossil fuels making the revolution possible by feeding the machinery, but the workers themselves needed food. The fertilizer that helped to produce it came from the islands of the Humboldt current in the form of guano, millions of tonnes of which were carried across the world, opening up one of the first import/export systems that spanned the globe.

But what is the actual source of all the nutrients? From where do they originate? All the productivity of the sea must take place in the upper lighted waters which act as a plankton-based solar cell, recharging the energy banks of the deeps. The output of the cell is determined by the amount of available nutrients present in the surface waters and its capacity by the depth to which light can penetrate. The two characteristics of the solar cell are irrevocably linked, for the greater the amount of nutrients the more plant plankton will be present, and the more plankton there are the less the light will be able to penetrate into the living water column–a perfect control valve.

If all the plankton present in the rich waters of the Humboldt current could be concentrated on the surface, they would form a green biscuit, a mere 0·2 millimetre (0·008 inch) thick. This may not seem much, but is enough to effectively screen off the life-giving rays of the sun and to feed the most productive fishery in the world.

Once the minerals are locked up in the members of the food web, their ultimate destination is down to the dark depths below. All those heads, feelers, mouth parts, arms and legs, chopped or sloughed off in the battle of life, and all the bodies of those organisms that are lucky enough to die of old age, are on a one-way trip to the bottom of the sea. There are, of

course, plenty of animals which may eat them on their way down and the more soluble parts may dissolve under pressure but their ultimate fate is in the hands of the force of gravity. Their destination is the ooze which covers the floor of the ocean deeps, a vast store of nutrients locked up out of recycling's way.

Now that is true of 99 per cent of all the oceans of the world but it just happens that the stretch of coast under consideration is part of the other 1 per cent. Here along the Humboldt current, upwellings stir the deeps bringing the nutrients in the ooze back into circulation in the light. Whether the upwelling is entirely due to the particular mix of wind, current and coastal topography or whether deep stirrings along the interface of the moving plates helps

Left Guanay Cormorants (*Phalacrocorax bougainvillii*), just part of a colony of more than a million resting after their own good haul of anchovies.

The plankton-based solar cell

White light is made up of all the colours of the rainbow and sea water acts as a filter which effectively screens off the light energy starting at the red end of the spectrum. Therefore, the deeper you go in the sea the less light there is to see by and the less colour there is to be seen. For instance, the lower of the two fishes in the lighted zone, appears much less brightly coloured, because some of the colours of the spectrum are missing at this depth. Even in the purest of sea water no light at all penetrates below a depth of about 1000 metres (3280 feet) and the enormous volume of water below this level is known as the abyss and remains at a frigid 4°C (39·2°F).

All the green plant life (phytoplankton) must therefore live in the upper lighted (euphotic) zone and together they act as a gigantic solar cell which feeds all the animals of the sea and supplies both them and the atmosphere above with oxygen.

euphotic zone the density of the plankton will, along with other factors, affect the depth to which the light can penetrate

abyssal zone

the sparser fauna of the deep sea include the anglerfish which catches its prey by means of a lighted lure

detrital ooze

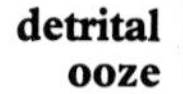

the fauna of the ocean bed include sea lilies and shrimps

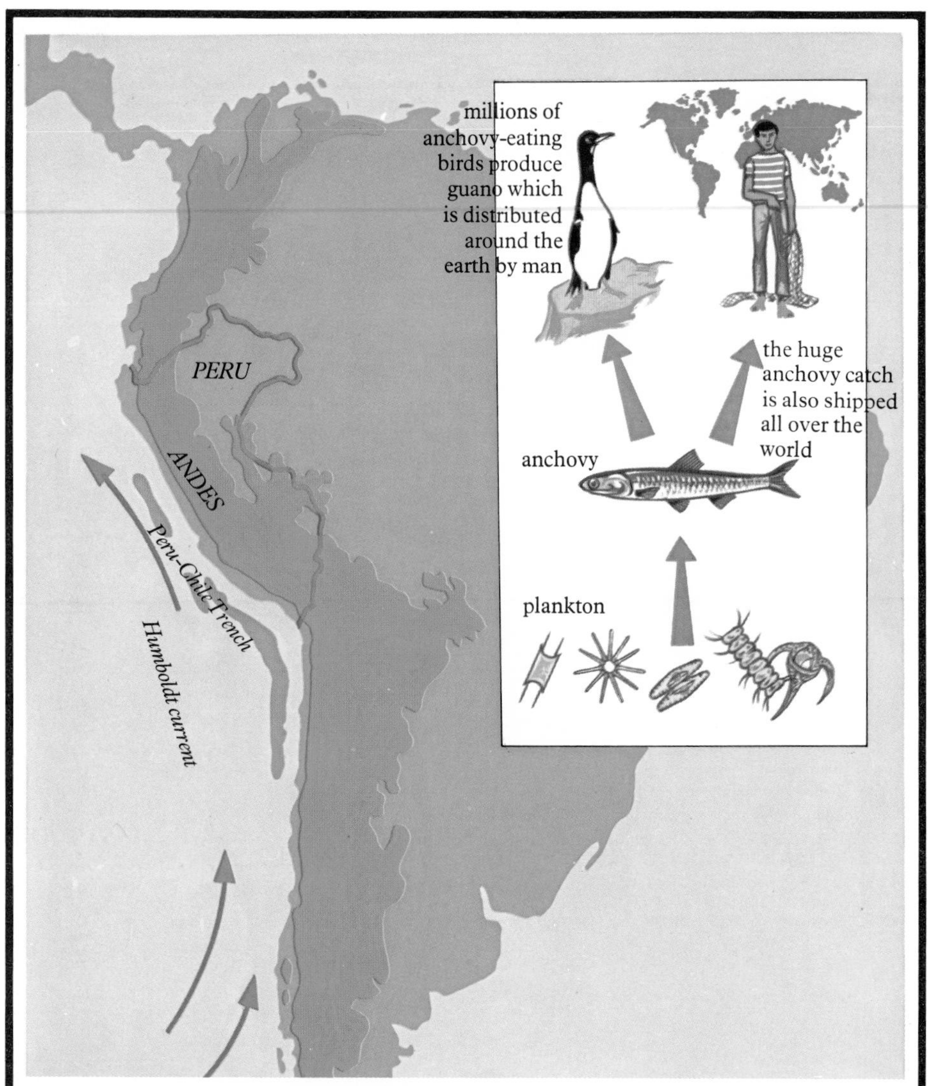

The end of a short chain

The enormous productivity of the Humboldt current has helped to feed the world of man. Anchovies are a good source of high-grade protein and as such we enjoy them, but mainly as a gourmet delicacy. Such direct utilization makes good sense because it puts the anchovy eater only three steps along the energy chain from the sun. The same goes for the Guanay Cormorant (*Phalacrocorax bougainvillii*) and the other birds which inhabit the guano islands off the coast of Peru. Guano is an excellent fertilizer and again it seems sensible to make use of the waste from the birds in order to grow more food for us humans. However, in recent years the productivity of the system has been declining. Perhaps the problem came when man tried to short cut the chain by using the fish directly as a fertilizer. Cutting out the middle man led to a more abundant and cheaper end product, and perhaps to overexploitation. Only the research which is going on at the moment will tell us why there are today fewer birds on the islands and fewer anchovies to be caught.

in any way, I do not know. However, the nutrients are recycled and this, together with the two-step food chain, makes it the most productive marine system on earth.

In recent years the anchovies have not been coming up to the expectations of the world's largest fishing fleet. Accusing questions have been pointed in many directions–is it overfishing, pollution, a warm counter current from the tropics, or has the mechanism that brings about the upwellings ceased to function? It could even be that a two-step food chain which lacks the diversity of a back-up team, and hence the ability to switch energy channels, is inherently unstable. A warning, perhaps, for any society which invests its affluence in too few of its members. Only time will tell. Evolution has taken some 3 000 million years to perfect the system that supports the anchovies, and man has only had 100 000 to perfect his.

So it was that evolution conquered the main problems of ocean life in a flurry of adaptive activity, which laid all the foundations for the future.

Evolution did not just happen in the open ocean. At the same time, it was beginning to reap the benefits of the inshore waters where, though light and gravity presented less of a problem, hanging on was much more difficult. The answer to the latter problem was in part solved by the seaweeds, which evolution furnished with an efficient adhesive holdfast. They were thus limited to growing on exposed rock which provides a clean surface, and none of the large members of this diverse and beautiful group ever conquered the shifting sands and silts of the inshore environment.

Success may be measured in a number of ways, not the least being stature, and as the seaweeds number among their ranks the largest annual plant in the world this is a fitting measure of their success. *Macrocytis pyrifera* var. *Humboldti*, the giant kelp, forms underwater forests and its fronds may grow as much as 30 metres (100 feet) in one year. As its last name implies, this particular variety grows in the cold waters of the Humboldt current. The sub-aqua forest, like its dry-land counterpart, forms a habitat for many other forms of life including other smaller seaweeds, red, brown and green, their predominant colour fitting them to particular light environments within the range of depths of the forest.

The seaweeds have evolved to be masters of the rough tough environment that fringes the dry earth. Their success depends in no small part on an efficient, though often complex, life cycle which includes at least one stage where free living cells must swim through the water. Apart from this hiatus of reproduction, which reveals the dual personality of many, they stand buoyed up in a solution of all the raw materials they need for healthy growth. Their delicate fronds are brought to the energy of the sun at each low tide and their sporelings compete for a place on the crowded rocks of the beach. Yet there they were destined to stay, for it was left to other organisms to take evolution's most difficult step up on to the land.

Before we leave the sea brimming with life I want to be self indulgent and introduce you to my favourite

The largest annual in the world

Algae is the composite name given to the members of a number of groups of plants, the vast majority of which are photosynthetic. Their plant bodies are not divided into stem, root and leaf and do not contain vascular tissue. The term seaweed is usually reserved for the larger members of the group which make their home in the marine environment. The most familiar seaweeds of temperate waters are members of the Phaeophyta (brown algae) the wracks, tangles and kelps, which dominate the zones between and just below the tides. In the tropics the most abundant seaweeds are the calcareous red (Rhodophyta) and greens (Chlorophyta) which play an important part along with the corals in making up the substance of the reef.

As all the seaweeds spend most of their life in water there is no problem with the movement of their motile reproductive spores. At high tide these are washed from the fronds of the sporophyte. The gonozoospores, as they are termed, settle on the seabed and give rise to either male or female gametophytes. The male gametophyte produces flagellated sperms which swim to the female gametophyte which in turn produces oogonia, each with one egg. After fertilization the resulting embryo divides and grows to form a new sporophyte.

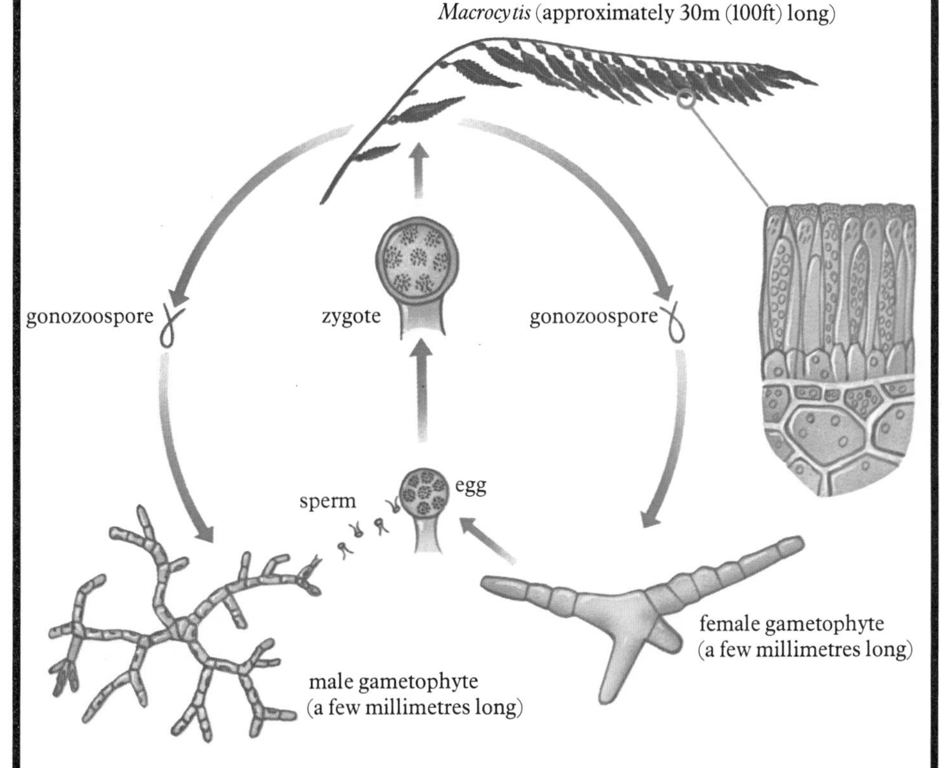

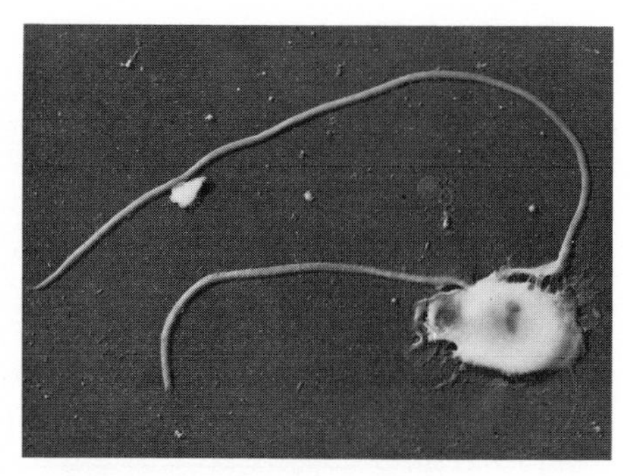

The male gamete (sperm) of the brown seaweed *Fucus* as revealed by a shadowcast direct preparation under the resolving power of an electron microscope (× 6000). The large white body contains the genetic information and the message is sped on its way by the beating of two flagella. The longer (only part of which is shown) is of the smooth whiplash type, while the shorter one is covered with what looks like tiny hairs (mastigonemes) and is referred to as the tinsel type. The body of the sperm must also contain all the chemical information and mechanisms for releasing the energy to activate the flagella. Although the electron microscope allows a lot of detail to be seen, you are actually looking at a picture of a dead, highly manipulated cell, which may have been much altered by preparation.

fossils, the trilobites. My favouritism stems from the fact that the first fossil I ever found was of one of these endearing creatures. They had their heyday in the late Cambrian when they were represented by at least 700 genera. They had a good run for their money, existing in a variety of forms for some 300 million years, and some of the forms, with enormous globular eyes each with over 1 000 lens-like segments, could have seen the first stages of the colonization of the land. If only some had survived until today–but then, of course, we would never have seen whatever it was that took their place. For at least 100 million years they ruled the seas and perhaps that is enough success for anyone.

Trilobites

A small sample of a fantastic range of fossils. The trilobites were the first arthropods (joint-legged animals) to rule the seas. Their bodies consisted of a head shield which was adorned with sensory appendages and bore the eyes; the thorax which consisted of numerous segments each of which bore a pair of legs; and a tail shield. Each limb was a general-purpose appendage with parts for walking, swimming and breathing and a paddle for wafting food forwards towards the mouth. The whole of the body and its appendages were covered with a tough skeleton of chitin, which is the reason that they produced such good fossils.

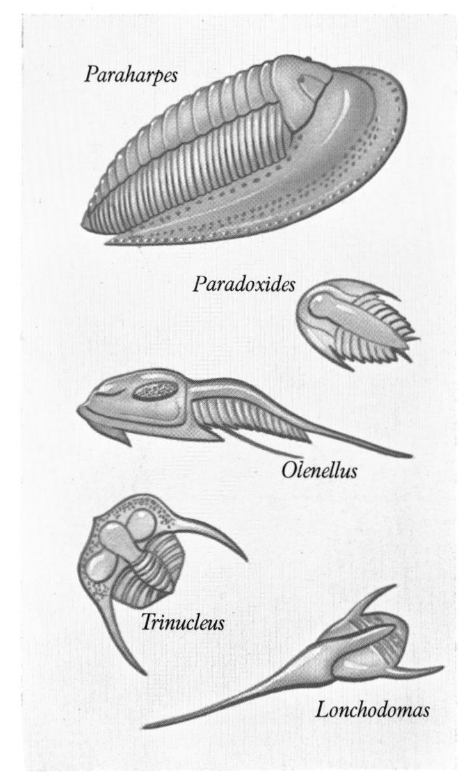

Chapter 4

Potential energy

How did evolution accomplish her most difficult task of moving from the sheltered environments of the sea to take up the challenge and opportunities of life on the dry land?

Because of the special conditions which existed in the tropical estuaries of the world, evolution dragged itself by its own rootstraps up above the reach of the highest tides and in so doing solved the problems of life on the land and laid down a vast store of fossil fuel for the future.

Sitting beside a horseshoe crab with the warm waters of the South China Sea making ripples in the sand, I was all of a twitter. I was trying to get myself into the right frame of mind in which to make the journey of my life; a journey that would take me back, at least in my imagination, into that part of the past when there were no plants or animals that could live on the land.

My companion on the shore should really be called *Limulus polyphenus*, for it is not a crab at all, but a more primitive sort of arthropod (that is, related to the spiders and scorpions). *Limulus* is one genus of animals that really does deserve the name a 'living fossil'. Its ancestry can be traced quite clearly back into the Cambrian and its larvae bear a remarkable resemblance to a trilobite. What better companion for a trip back through time, a trip down its own fossil 'memory lane'.

However, this trip was not to be entirely a figment of my imagination for Borneo offers two glimpses of what the earth looked like back in the early days of the land plants. One was quite close, behind the mangroves that fringe the coast at this point; the other was inland.

If you say Borneo to a biologist it immediately fires off three images in his mind–Orang-utan (*Pongo pygmaeus*), the great ape to which I must admit with some pride I bear a passing resemblance; *Rafflesia*, the parasitic flowering plant which boasts the largest flower in the world; and the saliva of *Collocalia fuciphaga* which the cave swiftlet uses to make its nest and we use to make bird's nest soup.

The genuine bird's nest soup bird (it says so on the packet) builds its nests in large caves that have been eroded into a gigantic block of limestone that forms a prominent feature of the north-east of Borneo. Beyond the caves is a second limestone massif called Gunong Api which is, if anything, much more impressive but about which very little is known. It rises up out of the steamy heat of the jungle to peak at around 1680 metres (5600 feet), its flanks covered with a mixture of scrub forest and what is best called tropical upland heath. It is, without doubt, one of the most inaccessible parts of the world and to prove it, while I am writing this

Preceding page Pongo pygmaeus, the Orang-utan. The large cheek pads show that this is an adult male. They usually live apart from the small social group of about six animals, which centres around the mother and latest infant. Their staple food is fruit with which their home, the forests of Borneo, abound.

Above *Rafflesia*, the plant which boasts the largest flower in the world and a most revolting smell which attracts insect pollinators and botanists alike, is not uncommon in the forests on limestone in Borneo. It is a parasite living on the sap of vines. Although the flowers may be up to 1 metre (3·3 feet) wide, they only last for a few days.

chapter, the first major scientific expedition to visit the area is camped on its dizzy heights. One of the objects of their enterprise is to study the most extreme form of karst landscape that exists. I reckon that it ranks with the wonders of the world, but only their work will tell. Fine pinnacles and knife edges of white limestone sawtooth up for as much as 60 metres (200 feet) above the heathland and between them rivers lose themselves in deep sink holes, each of which is rimmed with dense forest.

There is little doubt that this extreme form of karstscape is the result of the particular environmental cocktail that exists there at this time; a mix of erodable limestone, some of which is softer than other, an equitable temperature and torrential rain. Whether the vegetation helps in any way is not clear, although the fact that in places it hangs around the base of the pinnacles like a schoolboy's sock may indicate that it plays some part in the process. All living organisms produce waste as a by-product of their respiration and this includes hydrogen ions, electrons and carbon dioxide, all of which help to produce weak acids. Acid and limestone do not mix or when they do, they do so with a fizz as the limestone dissolves and carbon dioxide is released.

Whatever the exact mechanism the effect is the same–bare eroding rock–and this is the most spectacular place I know where you can look up at a landscape and say there should be vegetation there, but there is not. The annual march of temperature is almost perfect for plant growth, there is plenty of rain, but the pinnacles are devoid of life. It is a warm, very wet desert. So was much of the land around 450 million years ago. Although much of the centre of the land was probably a real desert, devoid of water, around the edges, and especially close to the coasts that felt the effects of onshore winds, there must have been rain in plenty. However, as there were no plants and animals capable of living on the land, it was devoid of life; the potential was there, but there was nothing to exploit it.

Now that is an easy statement to make but a very difficult one to prove because if anything was growing there, the likelihood of its being fossilized is very slim. We also know that the primitive plants, blue-green algae, bacteria and so on, had been thriving in the sea for a long time, and we know that their contemporary descendants are more than capable of growing on bare rock. In fact, a close look at the karst pinnacles of Borneo reveals the presence of a tell-tale blue-green jelly. So they may have been there, and if you want to see what they look like then go out to your own backyard, after rain, get down on your knees and take a really close look, because I bet they are growing there as well.

If blue-green algae were present over large areas of the wet coastal hinterland of the earth, then they would have been performing a number of very important roles. Firstly, they would have helped to put free oxygen into the atmosphere and the importance of that can not be overstressed. Oxygen, as we have seen, is a very reactive element and its presence in the contemporary atmosphere is entirely due to its continued production by green plants in the light. If the contemporary supply were stopped, our oxygen reserves would last a mere 2000 years.

Secondly, they would have helped to take nitrogen out of the atmosphere. Nitrogen, in direct contrast to oxygen, is a very unreactive gas and that is one reason why it is an abundant constituent of the atmosphere, 79 per cent to be exact. As nitrogen is an important constituent of the life chemicals the abundance of nitrogen can not be bad. Unfortunately, in its free state nitrogen is unavailable and before it can be taken up and used by plants it must be turned into nitrate (NO_3), a process which uses up a lot of oxygen. Even more unfortunate is the fact that the majority of nitrates are very soluble, so once formed they are rapidly washed away by the rain. Some of the earliest forms of life overcame the problem by evolving their own internal fertilizer factory, a chemical mechanism which could fix atmospheric nitrogen, producing nitrates, some of which could be used by the cell and some of which could leak out. This produced a ready made supply for others, a job which they still do today. Important things are they not!

The third achievement of these highly accomplished organisms relates to a third component of the atmosphere, and one that is present in comparatively small amounts–the gas carbon dioxide (CO_2). It is a raw material of photosynthesis, a product of respiration and, as carbon is the backbone constituent of all life chemicals, its delicate balance in the atmosphere is of great importance to all living things. All plants, however large or small, are therefore prime movers in the cycle that keeps carbon on the go.

Standing gazing up at almost bare pinnacles of rock it is sobering to think of the important role those tiny plants have played in the three great cycles of oxygen, nitrogen and carbon dioxide throughout evolutionary time.

There is, of course, one more cycle in which life plays a small though very significant part, and that is the incessant cycle of water. The life in the sea plays no part in the cycle itself, it simply makes use of the main reservoir of water which happens to be far from pure. Evaporation releases the pure water from the bondage of the salty sea and, after a period of riding high in the atmosphere as water vapour, it returns under the force of gravity as sweet (almost pure) rain, some of which falls on the dry land.

It would be true to say that, until land life evolved, the landward-bound water was a catalyst without a living cause. The blue-green algae were among the first living reservoirs, each cell holding a minute amount of water in store against the pull of gravity. From such small beginnings, the vegetation of the dry earth increased, sometimes with spectacular leaps and bounds, up to its present guesstimated figure of 13×10^{14} litres ($2{\cdot}8 \times 10^{14}$) gallons of water held in structured living form against the pull of gravity.

Not only does life hold on to water but in so doing it protects the earth on which it grows from the erosive power of the rain and it is clear that for a long time the whole of the surface of the land must have been exposed to all the forces of erosion. Just like the karst pinnacles of Gunong Api, some of the land had enormous potential for life with

Cycles the blue-green algae set in motion

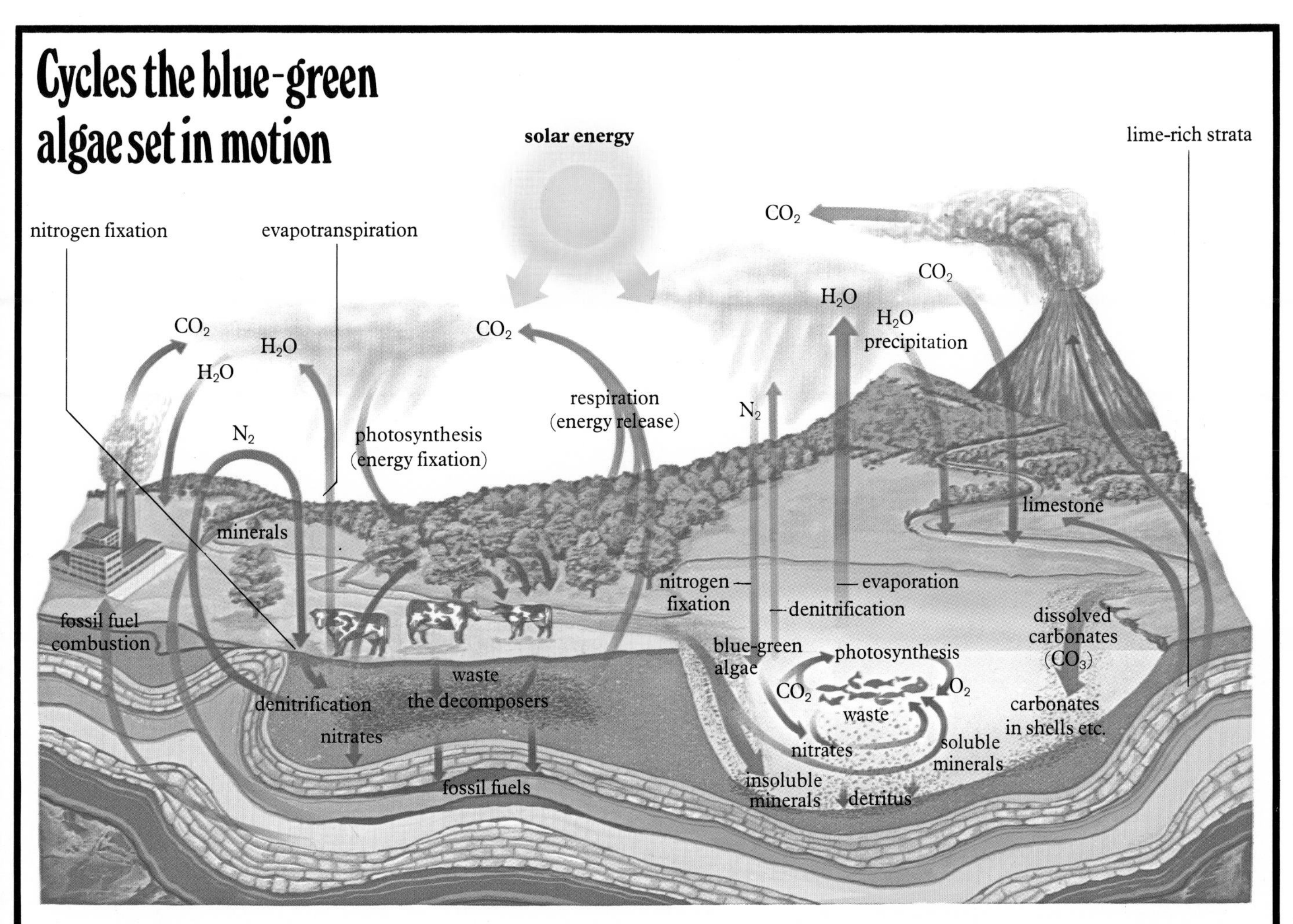

From small beginnings the major cycles of the living world have built up to their present levels, all of which are given below in 10^9 tonnes (1 tonne = 0·98 ton).

Carbon Every year photosynthesis uses up and respiration replaces some 50 units of carbon dioxide. The main reserves are 50 million units in 'fossil' form, 10 000 of them in coal and oil and the rest in the form of carbonates (CO_3) in limestones and other sedimentary rocks; 40 000 in the sea; and a mere 800 in the atmosphere. The fact that all the 'fossil' reserves are the result of past ages of photosynthetic affluence shows that the annual carbon budget has not always been in balance.

Oxygen is the most abundant element in the earth's crust accounting for around 50 per cent of the total. Its circulation is irrevocably linked with that of carbon, the amount of oxygen being something over double that of carbon. An important reserve of oxygen is that which is combined in the sulphate (SO_4) of both sea water and of sediments.

Nitrogen is at present being fixed into the biosphere at the rate of 0·092 units per year. The corresponding loss to denitrification is only 0·083 units. If these figures are correct then the biosphere is at present in positive nitrogen balance.

Water. To produce the annual total world crop of 115 dry units, it requires some 45,000 units of water, only 67·5 units of which are used as a raw material for photosynthesis, the rest being lost in transpiration.

(From 'The Biosphere' by G. Evelyn Hutchinson.)

Water, water everywhere, and the only drops we can drink are those 285 cubic kilometres that fall on the land surface and we have to catch what we need while it flows back down to the sea. The same goes for all the plants and animals that have evolved to live on the dry land.

The water cycle

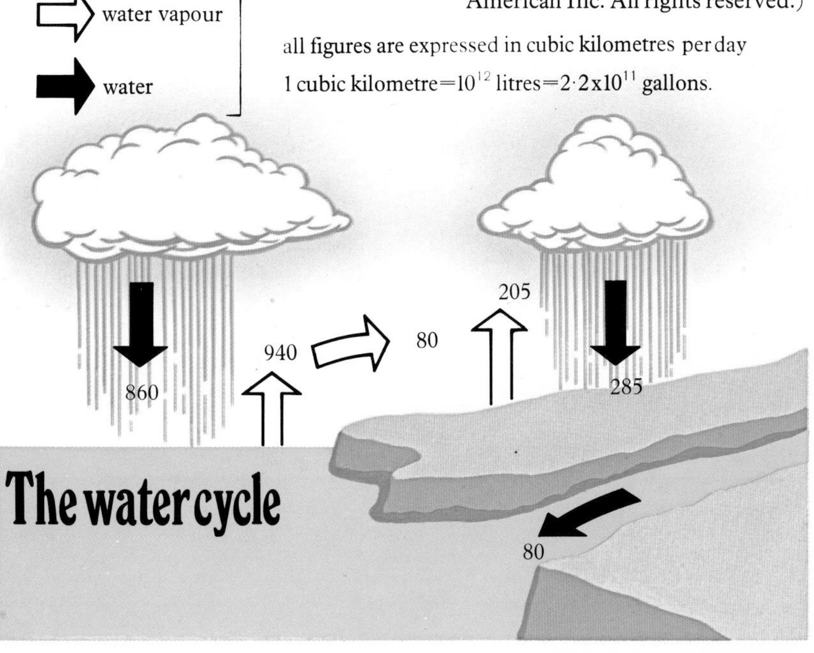

Overleaf Waves crashing on the shore may make a beautiful picture but they do not produce a very comfortable habitat. How would you like to have to live permanently between the tides? It is tough at the top and it is not all that much better at low-water mark.

The kelp, *Laminaria digitata*. The tide is out revealing the branched holdfast which anchors it to the rock and the fact that the stipe has enough strength to stand semi-erect.

nothing to exploit it, and erosion by water was rife. The ultimate fate of the products of that erosion was to be carried down to the sea but, unlike the water, they were destined to stay there for a long time, and thereby hangs the next part of the tale.

There is nothing to suggest that the land plants arose directly on the land and it is difficult to imagine how the necessary changes could have been accomplished in such a harsh, unstable environment. All the evidence suggests that the ancestral stocks of the land-livers arose out of the sea, but again it is difficult to imagine how all the necessary steps could have been taken, especially when you remember that the major step had to be taken across one of the harshest environments that exists. The littoral zone, the land between the tides, does not at first sight seem the best place for evolution to take place–crashing waves, cold and wet one moment, hot and drying the next or, in the depths of winter, warm and wet alternating with drying and freezing. Yet the littoral zone did have one distinct advantage. On the shore, the major changes of environment, the rise and fall of the tides, occur at regular intervals and they have done ever since the moon went into orbit around the earth. This means regular bath times and regular periods in between, and the actual time of immersion depends on the position on the shore and it varies with the phases of the moon on a two week cycle of extra high and extra low tides. Perhaps it is not such a bad place to test out the land legs of evolution.

There is one environment between the tides where the changes are not so abrupt. In the estuaries of all rivers, great and small, the sweet and salt waters mix and mingle to produce a series of zones, from salty through not so salty to hardly any salt at all. The constant erosion of the land provides sediment in plenty, which may be deposited in the estuary, building the substratum up above the effects of the swirling sea waters. If there is an optimum place in which to sort out the problems and make the transition from wet to dry and from salt to sweet, this is it.

By this time my companion of the estuarine silts had taken himself off back into the safety of deeper water and so I started out on my journey alone. Skirting along the empty beach I sorted through the flotsam on the strand line until I found a piece of brown seaweed, nothing on such a grand scale as the giant kelp of

Peru but a piece of *Sargassum* about a metre (3·3 feet) in length. As seaweeds go, *Sargassum* is pretty tough, able to fend for itself floating on the Sargasso Sea. It can also live in warm tropical waters which are shunned by most of the other large brown seaweeds, but it can not live above the tide mark. Although it could not have been out of water for more than six hours it was already drying, the more delicate leaf-like parts of its body (thallus) were shrivelled and cracked beyond recognition. The word thallus means a plant body that is not divided into root, stem and leaf, and, as we shall see, without such a division the seaweeds were destined to stay in the sea.

What is it that a land plant needs that a seaweed has not got? First and foremost it needs a protective coat to help keep the water in, but it must be a very special coat because it must also let the light shine through and allow carbon dioxide to diffuse in efficiently enough for photosynthesis to take place. All in all a fantastic design problem, and that is just a start.

When the tide is in all large seaweeds stand to attention, floating upright in the water. They are thus formed into a structured 'forest' system composed of several different layers, the 'canopy', 'understorey' and 'ground layer' receiving differing amounts of light and protection from the waves. On land the plants would not be so lucky, and a whole lot of seaweed-like plants lying about on the surface of the land would not be much of an advance or an advantage over the blue-green jelly mob. To be successful on the land, plants require a tissue or tissues which can support them in the non-buoyant air, allowing them to get up above their neighbours forming a structured forest system in which there are many environments providing places in which a lot of different plants can grow. Not a single super organism but a complex super system, providing potential for a whole range of evolutionary endeavour.

Sea water not only supports the seaweeds in a physical sense, it also provides them with all the water and nutrients they require. The main function of their holdfasts is to hang on tight to the rocks, from which they extract directly few, if any, of the minerals they need for growth. In contrast, a land plant needs a rooting system which can grow down into the soil, anchoring the plant and at the same time tapping the necessary supplies of water and minerals.

So, the land plant needs roots to collect water and minerals and stems to support the plant body in a more or less upright position, holding the leaves up to the sun. Each part of the plant does a particular job which creates a problem–the need for an effective transport system to connect them all up. Water and salts must be transported up from the roots to the leaves and energy rich sugars back down and around, to keep the whole plant growing.

Last, but by no means least, there is the problem of reproduction. In the absence of free water how can the 'sperm' swim to fertilize the egg and

Overleaf Assorted seaweeds, green, red and brown. The tide is out and they are all at the mercy of the wind and the sun.

One foot in the water

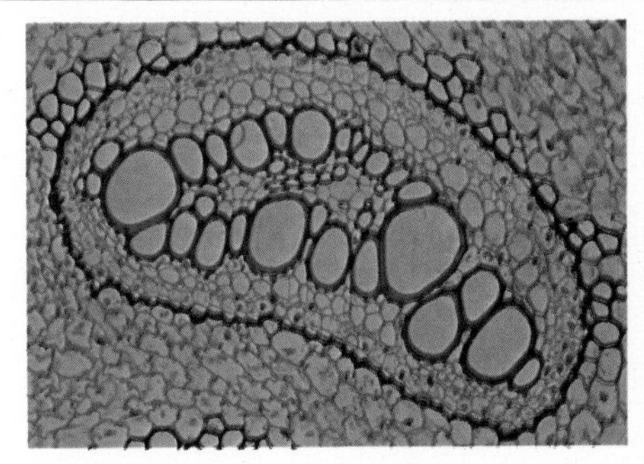

Part of a section of the stem of Bracken, *Pteridium aquilinum*, a modern pteridophyte. The red tissue is the xylem or wood which gives strength to the stem and acts in the transport of water. The surrounding thinner-walled green tissue is the phloem, the function of which is to conduct sugars and other manufactured substances around the plant. Together they constitute the stele or vascular tissue bounded by a layer called the endodermis. The stele is surrounded by a living ground tissue which makes up the bulk of the young stem.

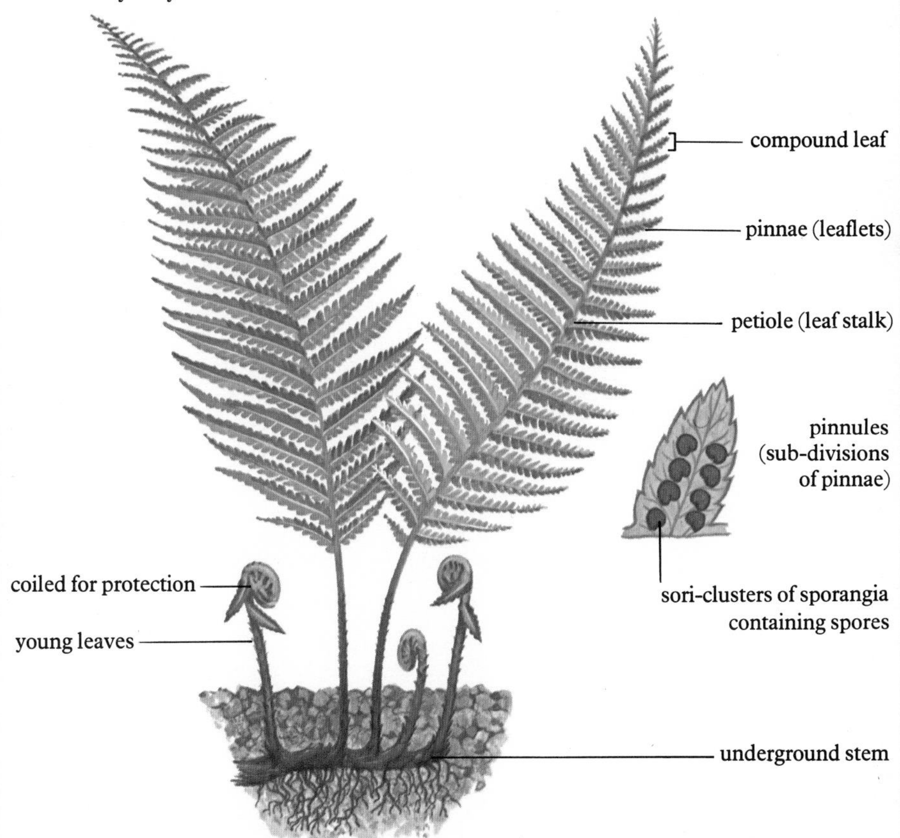

The ferns and their allies are said to be the first land plants. Check out their properties which fitted them for life on the land and their major hang up which left them with one foot in the water. Compare their structure and life cycle (p71) with that of the liverworts (p93) and note the advance in the size and complexity of the spore-bearing generation. It was the sporophyte which developed stomata and the ability to become homoiohydric by controlling water loss from its leaves. Without this ability the gamete-bearing generation was doomed to play a lesser role, at least in size, in the life of the plant.

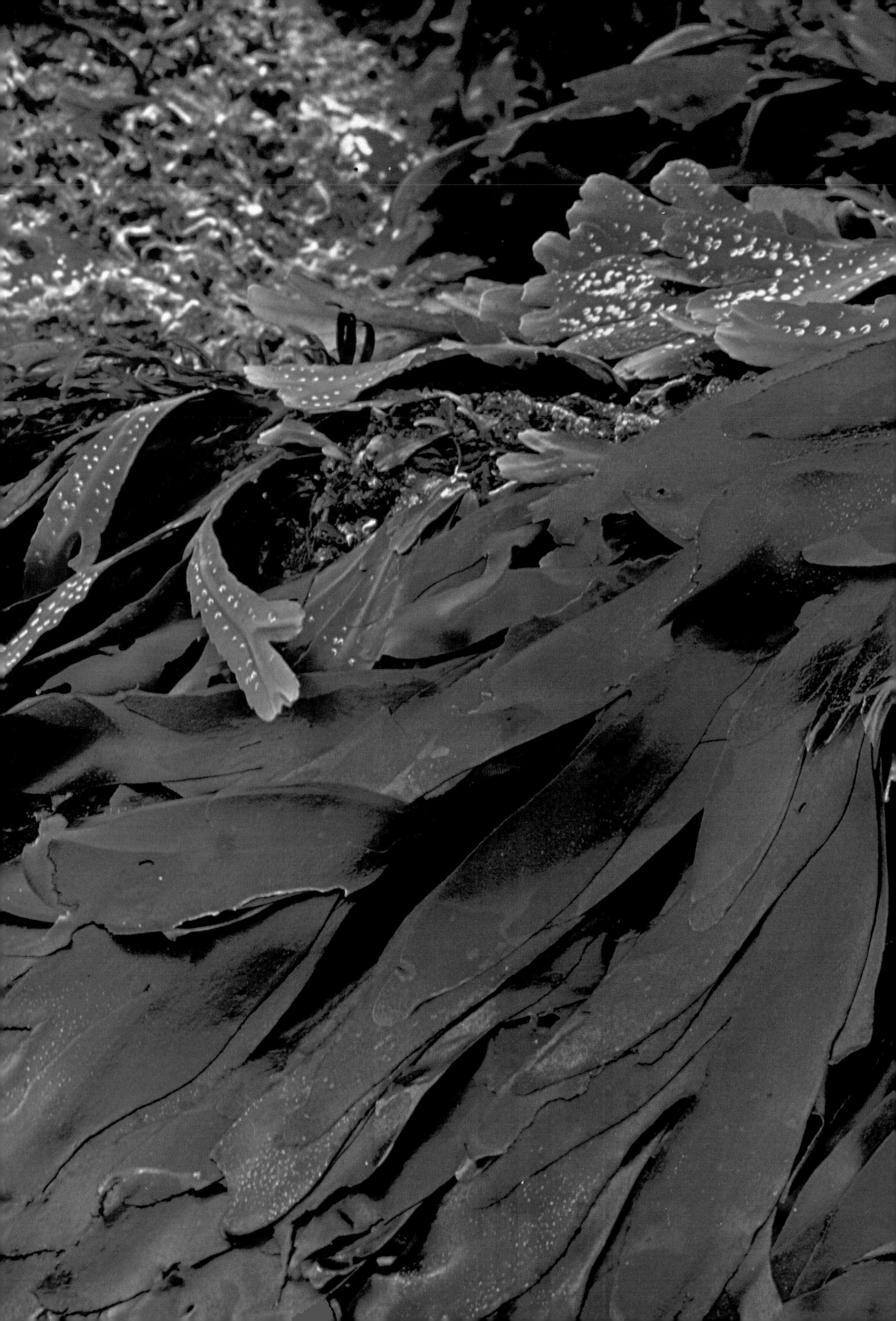

how can a mobile spore swim away from its parents dispersing the next generation? Added to this is a need to provide the young plant with a survival pack (energy store and protective coat) to tide it over any adverse conditions which might occur before it gets fully established.

Problems, problems, all of which had to be solved before plants could move on to the land. This was certainly going to be the most complex and most difficult step in the evolution of the plant kingdom.

Why should that step have been taken at all? The plants did not sit there in the sea and say to themselves, 'look at all that energy up there, we must get up and go'. Something must have caused it to happen and the answer was not far away from the point at which I stood holding my seaweed.

I made my way up from the beach and through the fringing mangrove. It was no good looking for the answer there, the mangroves are much more recent products of evolution. Further in, the mangroves began to be replaced by other trees, but again these were mere youngsters in the time scale of evolution. With the last of the mangroves I left behind the final traces of the effect of the tides and immediately the substratum changed from a sticky clinging clay to an equally clinging dark brown peat. If the going through the mangroves had been tough this was sheer murder.

I was now following an old trail that had been cut by engineers whose task had been to survey the terrain and, especially, to ascertain the depth of the peat. It was from their data that I knew I was heading in the right direction and that I was going uphill. My destination lay 4 kilometres (2·5 miles) ahead and 14 metres (46 feet) higher up, a slope of about 1 in 300, not really worth bothering about. Yet this slight slope was the main reason for my pilgrimage to Borneo, this was what I had come to see, for this is the modern-day counterpart of evolution's stepping stone up to the potential of the dry land. As I walked I got more and more excited–this was not just a stepping stone, it was a launch pad on which things really started to happen.

Imagine that you were a plant growing in a tropical estuary and that you and you alone had a strong support tissue that allowed you to stand up straight when the tide was out. Just think of the advantage you would have, overshadowing all those 'layabout' seaweeds–all that sun, all to yourself, enough to give you that get up and grow feeling, and that is exactly what it did to evolution.

The first real land plant that has been found in the fossil record lived about 400 million years ago and has been given the name *Cooksonia*. Rocks that were laid down over the next 4 million years, in places which are now as far apart as Scotland, Czechoslovakia, the United States and Australia, contain the fossils of a number of similar plants, all of which have been put into a group of their own called the Psilopsida.

They all had the following features in common–an underground rooting portion (the rhizome), and an upright stem that branched in such a way that each division produced two equal halves. This type of branching

The ferns and their allies

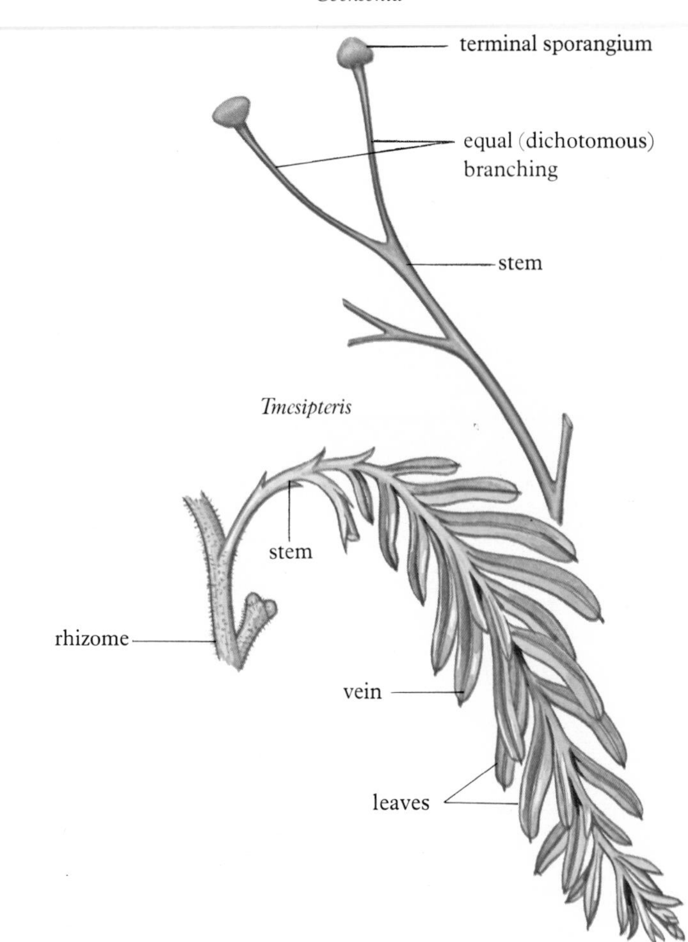

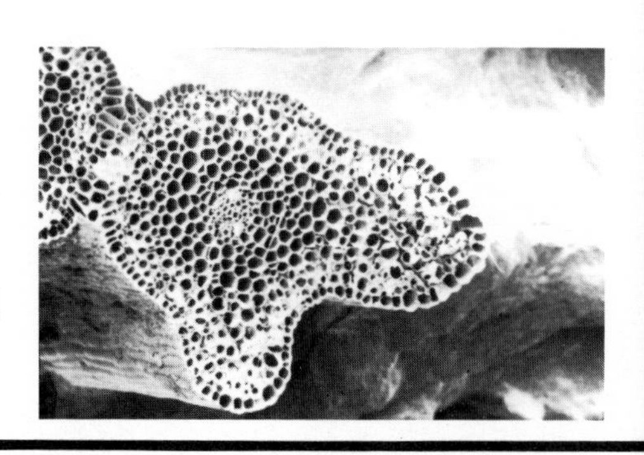
A section of *Psilotum*, a very primitive-looking stem. Below the epidermis, which is perforated in places by stomata, are several layers of cells which contain green chloroplasts. The stem is the main organ of photosynthesis. Below this is a zone of thick-walled strengthening tissue and below that a mass of thinner-walled ground tissue. The central vascular strand consists of an outer zone of phloem-like tissue with a central water-conducting xylem.

The reproductive sporangia are the only evidence that has ever been found of *Cooksonia*. It belonged to the family Rhyniaceae, a group of primitive plants that existed in the Upper Silurian and Lower Devonian.

The order Psilotales is extremely interesting, for it contains two genera – *Psilotum* and *Tmesipteris*, which have a level of organization very little higher than the earliest land plants and are considered close relatives. Yet these two primitive plants are alive and well today. *Tmesipteris* is found in New Zealand, Australia and Polynesia, *Psilotum* is widespread throughout the tropics and subtropics. The life cycle of *Psilotum* is shown in the illustration.

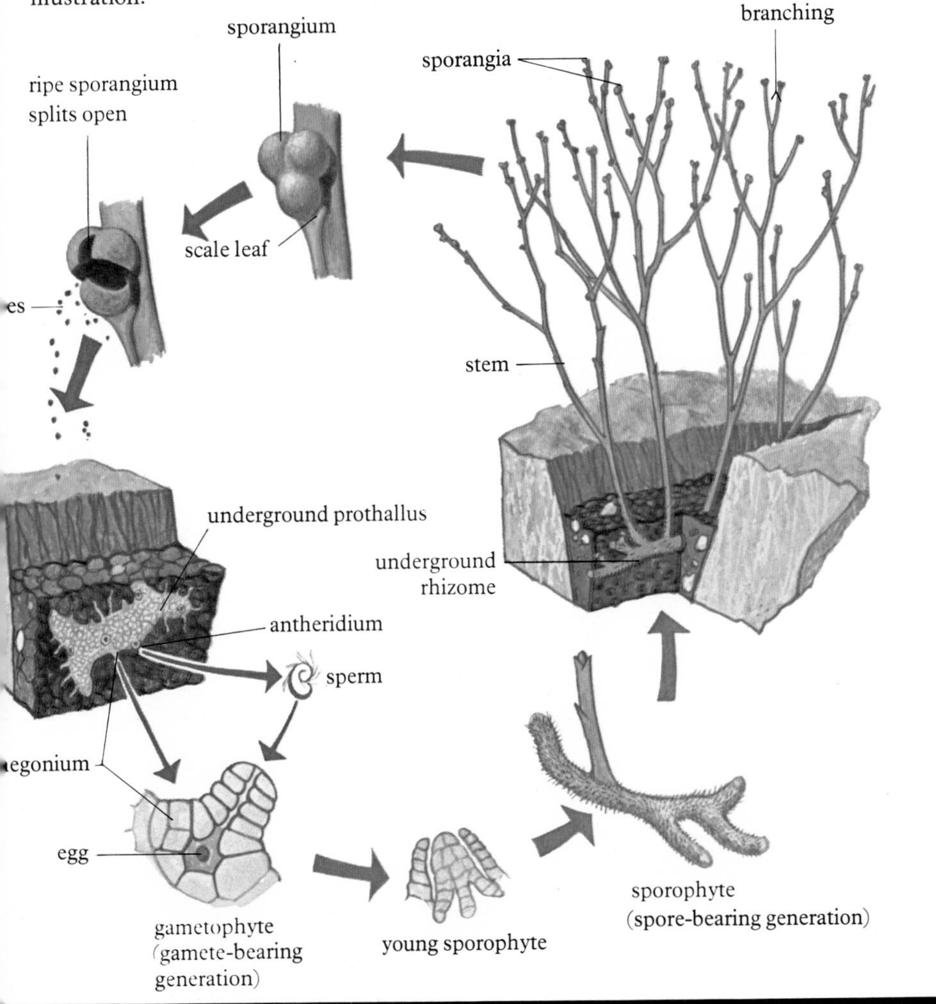

is called dichotomous and is very characteristic of the seaweeds. In some cases the fossils are so good that we can see that the cells of the outer layer of the stem were modified into a protective layer, the epidermis, which was punctured at intervals by pores called stomata. These are very like the stomata of all contemporary land plants, each pore being surrounded by two cells which can move, thus opening and closing the pore, controlling the ingress and egress of gases. When there is plenty of water in the soil, the stomata can open letting carbon dioxide in so that photosynthesis can take place. When water is in short supply the stomata can shut up the photosynthetic shop.

Each stem had a central strand of tissue (the xylem) made up of thick-walled tubular cells called tracheides, ideal for providing support and transporting water up the stem. Around the central xylem was a sheath of thin-walled tissue, some of which had sieve plates on their walls and thus bear a great resemblance to the phloem of modern plants. The phloem is the tissue which transports sugars and other manufactured substances around the plant body, the function of the sieve plates being, at least in part, to control the transport during periods of adverse climate.

A prominent feature of many of the fossils is the presence of terminal or lateral organs, which have a protective outer wall and contain many separate cells, each of which have a thick protective coat. Comparing these structures with those of modern-day forms it is safe to conclude that they are reproductive structures, sporangia containing spores, but further than this it becomes pure guesswork.

There seems little doubt that all these plants had many features which fitted them for life on the land and the evidence points to the fact that they were denizens of shallow peaty pools. Whether these were situated close to the coast or up river is more difficult to say. However, since the fossil record shows that for millions of years the only large plants had been seaweeds, it does not seem unreasonable to argue that these adaptations arose in the betwixt and between world of the marine fringe.

So all of an evolutionary 'sudden', primitive land plants, some over a metre (3·3 feet) tall, were found growing across a large section of the globe. Again, it would appear that once the main adaptive breakthroughs had been made there was no holding the process back. Perhaps even more interesting is the fact that once evolution had made the main breakthroughs–epidermis, stomata, xylem and phloem–it held on to them, for we find them in all the more advanced groups.

Unfortunately, none of the group of primitives made it through to the present day. There are, however, two genera of plants alive today that are looked upon as their closest relatives. One of them, *Psilotum nudum*, grows in the swamp forests of Borneo and, in fact, can be grown quite easily in a greenhouse if you have green fingers. It is from study of these modern forms that we can build up some sort of picture concerning the life cycle of the 'primitives'.

Psilotum grows as an epiphyte on forest trees, in cracks in rocks, and on

The 'advance' of the sporangia

The earliest forms of pteridophytes all bore their spores in simple sporangia which terminated the branches of each dichotomy. **1** Changes in the pattern of branching relegated the sporangia to a lateral position while extension of the stalk as a columella within the sporangium further complicated the picture, **2**. The presence of leaves led to the possibility of sporophylls bearing sporangia, **3**; at first arranged at intervals down the length of the plant, **4**; or in special cone-like structures, **5**; finally within such a cone the presence of two types of sporangia bearing two types of spores, microspores and megaspores, **6**. Faced with such evidence (and many intermediates) it is so easy to erect a series of evolutionary advance leading ever upwards towards the pollen grains and ovules of the conifers and flowering plants (see pp77 and 90–91). Whether they do in fact mark stages in an evolutionary path or simply different ways of doing the same job is still a matter of conjecture, although most of the evidence points to the latter.

1 *Zosterophyllum*
2 sterile columella *Horneophyton*
3 sporophyll *Protolepidodendron*
4 *Lycopodium selago* fertile leaf sterile leaf
5 cone *Lycopodium carolinianum*
6 microsporangia megasporangia *Selaginella*

sand, all of which are habitats that are subject to periodic drought. Spores released from the capsules are blown about by the wind and some will eventually arrive on a suitable damp substrate or they may lie dormant until it rains, protected by their thick coat. When the conditions are right they germinate and grow to produce a colourless plant which looks not unlike a short piece of the rhizome of the parent. They are, however, very different. Instead of growing up and producing chlorophyll and sporangia the tiny plants remain underground feeding saprophytically on decaying matter, and safe in the damp substrate they produce sex organs. The male organs are called antheridia, and each one contains many spermatozoids which, when released, swim about through the water in the soil, propelled along by many waving flagella. A single spermatozoid will fertilize an egg which is borne singly in a female organ, the archegonium. The fertilized egg will eventually grow into a new *Psilotum* plant, all green and gorgeous ready to bear spores and start the whole cycle off again.

Thus, it is easy to understand how the problems of reproduction on the land were first overcome. The group of plants which include not only

Psilotum and the primitives (what a good name for a group), but also the club mosses, horsetails, quillworts and the ferns are called the Pteridophytes. They are undoubtedly the first of the 'vascular plants', the plants with an internal transport system. They solved most of the problems of life on the land, but when it came to the problems of reproduction they side-stepped the issue by using a double life style.

The growth of the large spore-bearing 'extrovert' generation is no longer controlled by the presence of free surface water but can regulate its own way of life. In contrast, the smaller reproductive generation can only perform its main function of sexual reproduction in the presence of free water. This split personality or alternation of generations earned for them the title of the first land plants, bringing them almost unrivalled success for 65 million years – and were they successful!

During this time the pteridophytes flourished throughout the estuaries of the world producing trees of over 40 metres (132 feet) in height, giant lycopods and horsetails that formed structured forests within which many smaller plants could grow. They were so successful that in a roundabout way they gave their name to that particular period of geological time, for it was their remains, together with those of other evolving groups, that formed the great coal deposits of the world, hence the term the Carboniferous period.

Every time you burn a bucket of coal, you burn a bucket of history, for in the matrix of carbon are the hard coats of the spores which can tell the expert the story of how these amphibians among plants left their most delicate problem behind in the swamp waters.

I was by this time nearing my goal, the apex of a long slow gradient which holds the answers to many problems in a state of perfect preservation.

I was walking through tall tropical forest, nothing to be all that excited about. However, this is a very special forest, the nearest thing left on earth to the coal swamps of the past. No, I had not found the lost world. Unfortunately, I was not surrounded by giant lycopods and horsetails – they have all long since disappeared from the face of evolution. The actors were different but the system in which I was standing and in which they were playing their own particular roles was the same as that which existed way back in Carboniferous times.

The clue to the fact that this was no ordinary tropical forest was all around my feet. The dominant ground flora over much of the area were gigantic pitcher plants, which have the capability of catching and digesting insects. In fact, some of the pitchers of these particular plants were large enough to contain quite a large animal.

Why should a plant that has chlorophyll and the capability to

The Carboniferous period

Left A coal-forming swamp of the Carboniferous period. If only the primitives had not all become extinct, the swamp forests of Borneo might look something like this.

Right An energy sandwich – coal strata exposed by strip mining. Once the carbon in that hard black rock was carbon dioxide in the hot, humid air of the swamp forests of the Carboniferous period. Photosynthesis fixed it into plant material which then went to form soft brown peat. Time and the weight of sediment laid down on top turned it into coal. Very soon the fires of the twentieth century will turn it into carbon dioxide once more. I wonder where it will go in the next cycle?

photosynthesize develop an 'appetite' for insects, and why should it thrive? The short answer is that there are a lot of insects for it to capture (there is the potential). But what advantage can it be for a plant to become part of the consumer chain and a carnivore at that? Wherever we find insectivorous plants in abundance they are growing in acid, nutrient-poor habitats, where there is a distinct shortage of nitrogen. One is forced to conclude that a meal of insects must help to provide the plants with some of the nitrogen they require, short-cutting the normal cycle of growth and decay. Then what are they doing growing on the floor of a tropical rain forest where the cycle of growth and decay could not be more rapid? This is, however, no ordinary tropical forest, for when a tree falls to the ground it rapidly sinks in down below the ground-water level. Decay does start and very quickly reduces even the toughest trunk to a mush, but once below the ground-water table decomposition comes almost to a stop, for down there there is not enough oxygen to support a large population of decomposers. No decomposition means that the natural cycles of the forest are blocked and much of the nutrients are locked away forever. No wonder then that these plants eke out their meagre nutrient supply with a tasty insect or two.

The proof was below my feet, which were rapidly sinking into the forest floor. I selected a length of palm stem and pushed it down into the ground. It went in quite easily; certainly any fallen tree trunks must have gone through the softening-up process. Another palm and then another each pushed down on top of the last were rapidly swallowed up until with 18 metres (59 feet) of pole down I hit solid ground.

It was the writings of a forest botanist, with the appropriate name of J. E. Tenison-Woods, which first brought my attention to, as he called them, 'the coalfields of Borneo', for, given enough time and the right conditions, that 18 metres of peat could be changed and compressed to form coal. The fact that I could readily push a bamboo pole down through it signified that it had not yet undergone the transformation.

It was the detailed and painstaking

work of one of the world's greatest tropical forest botanists, J. A. R. Anderson, who first levelled and probed the depths of these deposits and demonstrated that these great forests of Borneo are growing on, and producing, a great store of energy and a detailed record of their own development. From his painstaking research, working under some of the most difficult terrain conditions in the world, I knew that I was standing on layer upon layer of organic matter. Each layer contains readily identifiable bits and pieces, twigs, leaves, spores, fruit, pollen, wing cases, legs and so on, of all the plants and animals that have lived and died on that spot over the past 4500 years.

All those plants had done their own particular thing, lived their own particular way of life, performed their own important role and in so doing had pulled the whole system up by its own rootstraps, up above the effect of the sea. In a mere 4500 years the swamp forests of Borneo have pushed back the tidal waters reclaiming 14660 square kilometres (5660 square miles) of land from the grip of the sea.

Just think what a similar series of forests could have done with 60 million years at their disposal. The pioneers of the Carboniferous period not only pushed back the sea but in so doing they pulled themselves and evolution up towards the full potential of the sun beaming down on the dry earth. It did not happen all at once, there were many setbacks and the whole process took a very long time, but a detailed record of how evolution moved up towards the dry land was left safe in the coal deposits of the world.

Safe? Well at least until some 300 million years later when a new product of evolution learned not only how to read the record but how to make use of the vast store of fossil fuel. Wherever there is potential?

Left Every pitcher tells a story. The end of a leaf modified to do a lot of different jobs – it it is green and has stomata so it can photosynthesize; the function of the pitcher is to trap and digest insects which are attracted in by the promise of nectar; the curl on the stalk is swollen and is inhabited by ants (you can just see their front door); the stalk of the pitcher is a tendril which the plant uses for climbing and support. The only parts of the modified leaf that do not live up to their appearance are the fangs, and the lid which does not close. The latter is a pity, because when it rains the pitcher can fill up with water which dilutes its digestive enzymes, and the hanging bath tub can become a convenient breeding ground for mosquitoes.

The 'coal swamps' of Borneo

Did it fall or was it pushed? In the case of the swamp forests of Borneo it was the latter. The anaerobic conditions present in the swamp waters led to the accrual of peat which 'pushed' the surface of the swamps up above the reach of the tides. Certainly similar ecological situations in the estuaries of the world some 400 million years ago could have provided the selective pressure needed to sort the homoiohydric from the poikilohydric in the 'race' to the potential of the dry land. Unfortunately, the swamp forests of Borneo are doomed to extinction. The only tree which can tolerate the rigours of life high on the dome of peat is *Shorea albida* and the members of this gigantic monoculture are showing no signs of regeneration. Perhaps it is too acid, for the pH of the high swamp waters often drops below 2; perhaps there are just not enough minerals to go round; perhaps the coal forests of the Carboniferous period faced similar problems and that is why they came to an end. These guesses may not be very scientific, but until we have more facts they are the best we have got.

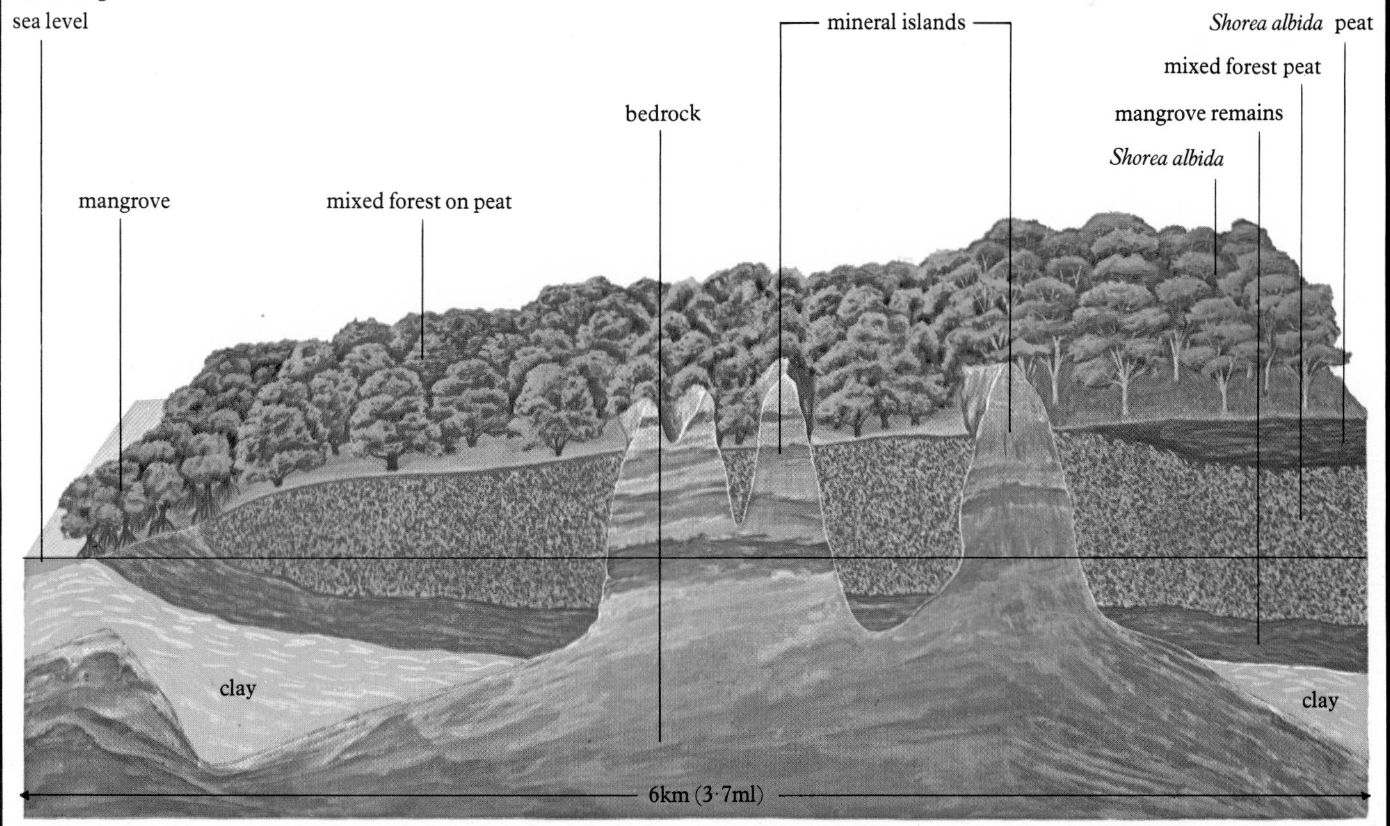

Chapter 5

Land of opportunity

If evolution had got no further than the amphibious ferns and their allies, life would have been stuck at the edge of the water; and if continental drift had not broken up the great land masses which existed a mere 200 million years ago, life would not have been able to penetrate far into their desert interiors.

But just as evolution was kitting out the cone bearers, flowering plants, mammals and birds (the real land plants and animals), so continental drift continued producing the land masses we know today, each one fit for a whole range of land life.

Three hundred million years ago, the sea was teeming with the full range of animals without backbones–protists, sponges, jellyfish, flatworms, roundworms, paddleworms, snails, starfish, shrimps, moss animals; you name it, by then evolution had produced it. Evolution had also gone to subaqua town with the chordates, which from small beginnings, like the sea-squirts and lancets, had produced an amazing variety of fish. The edges of the great continent Pangaea were clothed with forests of fern trees and horsetails growing close to the remnants of the swamp forests which now hid the great deposits of coal.

Conifers, although very like the giant pteridophytes, at least as far as their internal structure is concerned, had certainly taken a step further 'inland'. They have real roots and, as their name implies, conifers bear reproductive structures called cones which look not unlike those of the more advanced lycopods. Each one consists of a central axis around which modified spore-bearing leaves are clustered.

The cone leaves or sporophylls of the advanced lycopods have on their upper surfaces a micro- or a megasporangium which contains micro- or megaspores respectively. Microspores, once released, germinate to produce a male gametophyte which in turn develops male reproductive organs containing motile spermatozoids. Similarly, the megaspores produce female gametophytes with female organs, each one containing an egg. Fertilization is accomplished only after the 'sperm' have swum to the egg, and hence there is an amphibious stage in the life cycle.

In contrast, the microsporangia of the conifers produce microspores, called pollen grains, which are carried to the megaspore by wind. Once safe within the scales of the female cone, the pollen grain germinates to produce a long pollen tube which grows down towards the megaspore depositing the 'sperms' close to the egg. Safe within the cone scales each megaspore has developed into a female gametophyte complete with two archegonia, each containing one egg. The 'sperm' is thus delivered to the egg by the pollen tube, no need for flagella, no need for active swimming, no need for free water.

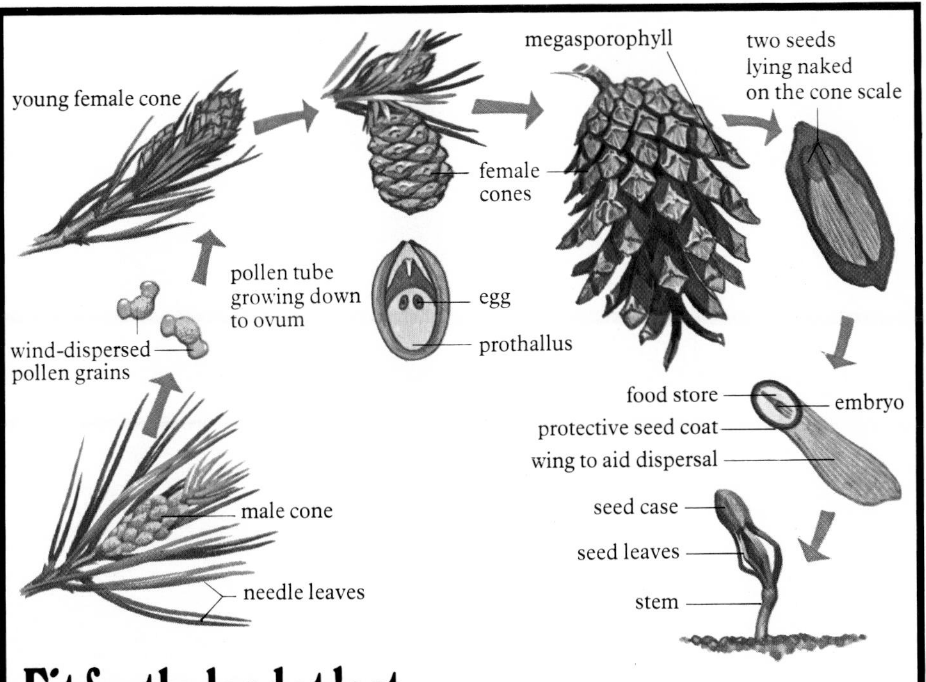

Fit for the land at last

The gymnosperms include more than 700 species, all of which are heterosporous, producing both micro- and megaspores.

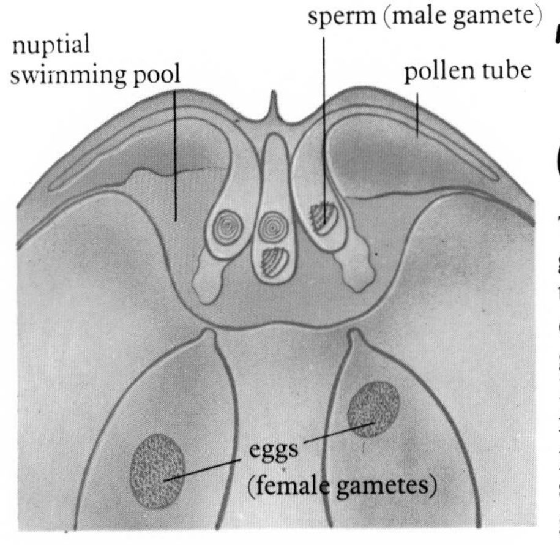

The last vestige of 'male lib'

The diagram shows all that is left of the gametophyte generation of the cycads. The tubes from three pollen grains have grown down and are about to release their male gametes into the nuptial swimming pool. If things go alright one sperm may fertilize each of the eggs which have been produced in archegonia nurtured by the sporophyte via the tissue of the female gametophyte.

Four cones of a modern cycad *Encephalartos altensteinii*, sitting in the centre of a whorl of large leaves.

The fertilized egg then begins development to produce an embryo complete with a food store, a tough protective jacket and a membranous wing to aid dispersal by the wind. The survival pack is called a seed. The conifers are, in comparison to the lycopods, 'single-minded fellows' that really look after the next generation. They are the first seed plants and the seed was their main key to life on dry land.

Again, exactly how all these changes, all these adaptations, occurred we may never know. All we can do is to look in the fossil record, comparing the records of the past with the successes of the present.

There are two fascinating groups of seed plants which appear to demonstrate what the half-way house may have been like. The first are the cycads which look not unlike short palm trees with gigantic and very tough leaves that spring from and help to form the short stout trunk. When they begin to reproduce there is no doubting the fact that they are conifers because some of them produce the largest cones in the world. Records like 40 kilograms (88 pounds) in weight and 60 centimetres (23 inches) in length are not to be sneezed at, although the amount of pollen produced by the male cones is enough to make anyone do just that.

With such large cones to deal with it is very easy to see that the cone scales are simply reduced and highly modified leaves. The pollen grains are carried to the female cones by the wind. Many will go to waste but some will fall on top of ripe megaspores which exude a viscous drop of fluid especially for the purpose of trapping the unsuspecting pollen grain. The fluid then dries up, dragging the pollen grains down into the ovule with it. Inside, the way has been prepared for the nuptials to commence. The tissue that overlays the gametophyte, which is already complete with two archegonia, breaks down to produce a fluid-filled cavity through which two large multiflagellate spermatozoids swim to fertilize the egg. This drop of fluid represents the last link with the world of free water, and the end of the free-living gametophyte generation.

Another plant which leads a very similar life style to the cycads is one that could at one time be described as the missing link. *Ginkgo biloba*, or the maidenhair tree, although very different in appearance from the seed ferns, with its handsome slender form and weird-shaped broad leaves and the fact that it does not produce massive cones, has a very similar cycle of reproduction complete with pollen tube, nuptial bath, mobile spermatozoids and all. This remarkable plant was known from the fossil record long before it was found growing in Chinese temple gardens and, much more recently, in the wild. It is, thus, the missing link that is no longer missing. The product of fertilization in both the cycads and the *Ginkgo* is a naked seed which contains an embryo and food store.

The seeds of these particular lines of success were sown around the edge of Pangaea about 300 million years ago, releasing the plant kingdom from the bondage of the swamp forests and other damp places, opening up the potential of real dry land at last.

The non-missing link, *Ginkgo*, is an offshoot from a group best called the Cordaitales which flourished for around 100 million years and probably gave rise to the modern conifers. The cycads, on the other hand, arose from a similarly important and widespread group that were contemporary with the Cordaitales and are best called the seed ferns. It is thought by some authorities that the seed ferns produced the stock which led up to the flowering plants of today. It is of great interest that both these mainstream lines of evolution went via a stage in which a mobile spermatozoid swam in its own private nuptial pool. This was the last stand of the male liberationists and the last stand of the gamete-bearing generation.

Onshore winds pushed cloudbanks over the margin of the great continent Pangaea feeding streams, lakes and rivers with a surfeit of sweet water for the growth of algae, mosses, liverworts and water ferns. Giant amphibians, some over 3 metres (10 feet) in length, went their ungainly ways, splashing through the shallows, seeking out their food. Smaller ones made short excursions on to the land, where they may have fed on insects. All of them had to return to the water in order to mate and lay their eggs. Like the trees beneath which they lived, these animal amphibians could only complete their life cycle in water–fertilization, spawn, to swimming larvae (cf tadpoles of living frogs) then through the metamorphosis of adolescence and on to carnivorous adulthood.

These first footers among the land vertebrates in part fed on the myriad insects, many of which lived a similar amphibious existence. As larvae they struggled to survive in the overcrowded world of fresh water, leaving it at metamorphosis for the freedom of flight paths in the dry air above the swamp.

It was a frontier world. The beginnings of real land life were there, an edging of vibrant activity, a green hem around the edge of Pangaea. The sea was overflowing with life but the centre of the continent was dead and would have so remained but for stirrings deep in the bowels of the earth.

Apart from the pteridophytes, insects and amphibians, other pioneers of the dry land were already in the making. The reptiles had evolved at some time in the past from the ancestral stocks of the amphibians and were leaving their marks on the dusty trails of the earth beneath primitive conifers that raised their leaves up to the sun.

Reptiles have dry skins, which not only help to shut in the all-important water but, unfortunately, help keep the oxygen-rich air out. Gas exchange between a reptile and the outside world is via the complex internal surface of a pair of lungs, aided by somewhat laboured breathing.

The early reptiles laid eggs, each equipped with its own leathery skin which helps to protect the baby lizard that lives safe within its own private developing tank. Fertilization is internal and, if comparison with their modern counterparts is valid, often goes hand in hand, or rather tail in

A land animal takes its first look at the world. Baby crocodiles hatch from their eggs which have protected them throughout development. In the various groups of reptiles it is possible to see a whole range of forms from those which lay eggs and then leave them to be hatched by the sun (oviparous) through to those in which the egg develops safe inside the parent and the offspring emerge ready hatched (ovoviviparous).

tail, with some pretty specialized courtship.

The earliest forms of reptile lived some 300 million years ago. They are called cotylosaurs and appear to be a half-way house between amphibians and reptiles. Much more lizard-like animals are found 100 million years later and are known as the Rhynchocephalia. One might imagine that all traces of these pioneers would have long since been swept off the face of evolution to become one of the top ten in the fossil record, but nothing could be further from the truth. The animal kingdom has not been outdone by *Ginkgo*, the last of the dry-land plant pioneers. Even today on some remote islands around that remote land mass, New Zealand, one member of the group still thrives in glorious isolation.

The name of this fabled beast is *Sphenodon punctatus* or Tuatara for short and his classificatory pedigree reads like something from the Royal College of Taxonomic Heraldry. The Tuatara may not be a top ten fossil, but he certainly ranks among the top ten animals that every budding biologist should know about and, if possible, should have seen in the flesh. I made my pilgrimage while on a round-the-world trip prior to writing this book.

My excitement at coming face to face with what must be the most famous of all living fossils was heightened by the helicopter trip out of windy Wellington with panoramic views of the islands and drowned valleys of Marborough Sound. The day was perfect and even the open sea was unruffled as the chopper slapped its way through the air,

chasing its own shadow over the parched pastures of Melville Island. Once a Maori stronghold, the island is now parcelled up into a number of farms, each populated by sleek stock which were sunning themselves in the heat of early autumn. There ahead of us were the sheer cliffs of Stephens Island, rising almost 300 metres (1 000 feet) out of an azure sea. The island itself was picked out by an untidy fringe of brown and white, a vital mixture of surf and giant seaweed braided by the currents into long flowing manes.

My head spun with the frail craft as we circled to touch down beside the stumpy lighthouse. A reception party was there in attendance–the families of the two keepers, hair bleached and arms bronzed by the combined effect of sun and salty winds, four dogs of varied shape and size, and a lady visitor who had spent her youth here completed the band of Stephens Islanders.

Sphenodon punctatus, the fabled Tuatara, emerges from its burrow, which was built by a prion, and looks out with two of its eyes on its modern island home.

Our guide, Don Newman, from the Wildlife Service led the way into the cool shade of the scrub forest that dominates parts of the island, and at once I felt very much at home. The reason was that in places the ground was covered by that well-known house plant *Tradescantia*, a recent immigrant to the island and already doing very well.

Don stopped in his tracks, bent down and deftly lifted into view a large male Tuatara. Annoyed by this rude elevation into the full glare of the sun flecks, the lizard opened its formidable mouth revealing that it closely resembles the 'beak' of a turtle, except for the presence of an array of triangular teeth. Those teeth are primitive character number one. Although very effective, they are no more than projections of the margin of the jaw and find no exact equivalent in the socketed teeth of higher animals. Click, click went the cameras, but there was no need for haste for this was no lone specimen. They were everywhere, small, medium and large (almost up to 60 centimetres (23 inches) in length); a bite from a big one would not only break the skin but your finger as well.

The secret of obtaining a good snap of (or from) a Tuatara is to find one with its business end sticking out into a sun fleck and then move in with great stealth. Tuataras, like other reptiles, see movement much better than they see objects and what is more they can feel movements through the vibrations in the ground. So if you want a photograph, it is slow motion all the way, one false move and your quarry will disappear down its burrow, or rather down a burrow which has been annexed from a seabird called a prion. The birds were at that time away at the other end of their migration route and while they are gone the reptiles rule the roost. At a rough guesstimate there are 20 000 Tuataras, all happy snappers. Once the birds return, a love-hate relationship ensues in which the Tuatara is not above devouring eggs, chicks or even a hapless adult as it returns gorged from its own feeding expedition out to sea.

Still it works and it may have worked for a long time, unchanged, but not that long, for there were

in the dampest spots where tadpole development is completed within the jelly-like mass. Finally, the froglets emerge to bolster this tiny population of 'rock hoppers'.

The ancestors of both *Leiopelma* and *Sphenodon* were once widespread around the living areas of Pangaea. So why, if they have stood the test of evolutionary time, surviving while evolution produced and cast aside the great dinosaurs, are they today found only in such a small area? What is the secret of their success in time yet contemporary restriction in space?

The answer lies in what, apart from the original cataclysm which ripped our particular ball of gas off from the embryo solar system, must be considered as the most important happening which shaped our planet for life. A happening that not only formed the world as we know it in our atlases but made it a fitter place for the evolution of life on land.

Around 200 million years before man walked upright on the face of the earth, Pangaea began to split up and the continents as we now know them started on the move; the pieces of the world's largest jigsaw puzzle began to go their own ponderous ways.

The land mass that is today New Zealand parted company from the main southern land mass, which has been romantically named Gondwanaland, about 160 million years ago. With it went the ancestors of the Tuatara, *Leiopelma*, and the modern-day fern allies and conifers. Whether the split occurred before the mammals and the flowering plants had evolved or whether New Zealand just was not lucky enough to take any of them with it, I am not sure. The evidence, however, points to the conclusion that New Zealand missed out in the share-out of evolutionary stock and went sailing away, an ark without its full complement on board. It also appears safe to conclude that the evolution of these new lines of advance, which would eventually dominate the dry earth, did not take place on the islands themselves. New Zealand had to wait for the flowering plants, the feathered birds and the hairy warm-blooded mammals to migrate in.

It is difficult to say exactly when the first flowering plants reached the Lands of the Long White Cloud. The fossil record puts it at around 65 million years BP (before present). Two flying mammals, bats driven across the South Pacific by storm winds, heralded the evolution of hair and mother's milk, joining the warm-blooded birds who had made the crossing some time earlier. Those same winds, though much later in time, blew Polynesian navigators in the same direction and 1300 years later brought Captain James Cook, the herald of everything which came with 'civilized' north-western ways. The mammals had come to stay.

At this time the Tuatara and possibly *Leiopelma* were much more widespread, living in all suitable habitats for they had no competitors and few enemies. Man smashed down the geographical barriers bringing with him dogs, rats, cats and a host of other mammals. With this massive new competition, the reign of the Tuatara on the mainland came to an end, its distribution gradually restricted to smaller and smaller islands to which the predatory mammals have not yet found their way.

There is, however, one spot on the mainland which remained inviolate by man for longer than most, in actual fact until 1836. Mirrored in the great Lake Taupo is a mountain massif on which ice and fire are mixed in a crucible of eternal steam and tortured rock. Three active volcanic peaks, Tongariro, Ngauruhoe and Ruapehu, jumbled craters, innumerable fumaroles and friable lava flows made this a forbidden area for the superstitious Maori. The first recorded human feet to tread the dreaded slopes were those of Charles Bidwill and party sometime in the late 1830s. This is a small part of what he recorded of the trip. 'As usual the men carried the children, and the women the potatoes and so on. The procession was closed by one or two pigs which, from the opposition they made to the efforts of the drivers, seemed to have as great a dread of Tongariro as the Maoris themselves. We were on Tongariro all day but the peak was never visible in consequence of the mist which covered its upper regions. The Natives, who were in a state of terror, maintained that the ascent of the mountain was impossible.'

Not far from its base the explorer passed the night with a fire made probably from two of the shrubs now bearing his name. The next day he made the first ascent. Charles Bidwill was the botanist who opened the eyes of the world to the alpine flora of New Zealand and I had always wanted to follow in his steps.

My first visit was on 26 March 1977 and I must admit that I did it the easy way elevated in a helicopter that was piloted by a Kiwi (the nickname for a native of New Zealand). His name was Otto Gram. The chromium rudder bars looked out of place beneath ten eager toes that played a continuous duet above the soles of his dusty flip-flops. He is the sort of guy you always hope to meet and are always very sorry to have to say goodbye to. He is a bundle of open good-humoured energy with an immense knowledge of his area. Those toes and his crushing handshake were simply extensions of his flimsy craft, newly fitted with a jet motor to give this symbiosis of man and machine the power they needed to ride the thermal wrath of Tongariro.

Up we went cradled in a perspex bubble, up and over the heather-clad slopes and on to the screes above. Otto indicated a large fault running diagonally beneath our flight path. 'Big one' he said 'it runs right through the mountain. I would have liked to have been around when that happened.' A sulphurous dew formed momentarily on the bubble as we rode above a line of steam vents and then we were skipping our way above three emerald green, or were they duck-egg blue, lakes. It was a scene of infinite calm beauty with a backdrop of hell itself. Behind the highest lake there rose a steep slope, stained yellow by sulphur and daubed with the red splash of a thousand fires, the whole alive with hissing steam.

We hovered upward and just where the curtain parted disappeared into the inner crater, into a funnel of lurid red-grey rock. We had entered the crater through a great gap that

Corybas oblongus is a ground-dwelling orchid with a really weird-shaped flower. There are between 15000 and 30000 different species of orchids and part of their success, especially in the tropics, is related to the fact that they are pollinated by insects. Due to this, many of their flowers look very much like the insect which acts as their own personal pollinator.

had been torn in one side of the cone by some gigantic force, the source of which gaped in front of us, a cavern that led down to the bowels of the mountain itself. Upward and out, we lifted into the fresh air to set down below this awesome spectacle of cataclysms past. Our task was to search for Bidwill's flowers which grow on what must be among the youngest rocks in the world and they were there, just as he had described–buttercups, eyebrights, gentians, pennycress; splashes of colour, the only signs of life in this young volcanic desert. What better place to think about the final phases of evolution.

The flowering plants or angiosperms are set above all other land plants by the possession of what are looked upon as advanced characteristics. Their water-conducting system is made up of a mixture of tracheides and vessels; the former are not unlike those of the ferns and the conifers whereas the latter are much more advanced. The vessels are, in fact, the height of evolutionary plumbing; having no end walls they form a continuous system of 'wide' pipes which offer little or no resistance to the passage of water.

The proper name for the conifers is the Gymnospermae, which is derived from two Greek words, *gymnos* meaning naked and *sperma* meaning seed, and the seeds of the conifers are just that, lying naked on the cone scales. The flowering plants, on the other hand, bear the name Angiospermae and the Greek word *angeion* means a vessel, hence seed in a vessel or hidden seed. Thus, the success of the flowering plants may be said to depend on two sorts of vessels, one to carry the water, the other to enclose the seed.

In the conifers we saw the last remnant of a real gamete-bearing generation. The small pad of tissue that bore the archegonia is all that remains of the separate free-living generation that, back in the algae, often vied in stature with the sporophyte. In the flowering plants even this last vestige of 'schizophrenia' has been swept away. The megaspore is an ovule contained within a special cavity, the ovary, that is formed by a carpel or sporophyll. At maturity the female gametophyte is no more than a single cell with eight nuclei that are the centre of stored genetic information. The pollen grains produce a tube which delivers two male nuclei (more information) to the ovule and fertilization is completed by a complex fusion of some of those information stores. The zygote, which contains the total chemical blueprint for a new sporophyte generation, develops into an embryo that is protected and nurtured within a seed, safe from the vagaries of the climate.

The microsporophylls (the anthers) and the megasporophylls (the carpels) are borne either separately or together in a flower protected by petals and sepals, which, like the scales of the conifer's cone, are made up of modified leaves. The parts of the flower may be structured and modified in a number of ways to expedite the process of pollination, and this development has put much of the detailed beauty into our landscapes. The fact that the most beautiful of the flowers are designed to attract insects, not humans, at least proves that we mammals do have tastes in common with the most advanced of that other successful line of evolution, the arthropods.

The first flower-bearing plants made their appearance in the Triassic period, when much of the available land was either dry desert or was already covered with a multitude of pteridophytes or gymnosperms. Great changes were, however, about to take place, for deep in the earth, some force that had remained quiescent for millions of years was beginning to tear the land mass apart. Areas which had been too remote from the sea to feel the effect of rain-bearing winds were now brought within the grip of the effects of the maritime climate. Many of the deserts of Pangaea could now turn green and the march of the land plants was in full swing.

Evolutionary fitness is compounded of two parts–the fitness of

The fabulous Kowai is a member of the important family of flowering plants once called the Leguminosae, now called the Fabaceae. It is an important family on two counts – firstly, many of its members have nitrogen-fixing bacteria living in special root nodules, which made possible the rejuvenation of farm soils by crop rotation; and secondly, the family has provided the world with important food plants like peas, beans and soya.

The flowering plants

Why have the flowering plants been the success story of the most recent geological periods? In essence the answer must lie in their adaptability, a statement which unfortunately does no more than push the question back by several steps. Perhaps the best way to try and answer it is to take a look at their special features and see if they give us or them any leads.

But first, why do we say that they are the most successful group in the plant kingdom? With at least 230 000 species in the field they are more than ten times as diverse as their oldest rivals, the algae; when it comes to productivity they top the bill (see the world vegetation map p122), as they do in stature (eucalyptus and lianas), although it would appear that the algae hold the length record for annuals (see p56). Perhaps their greatest achievement, however, is their range of habitat. On land they are found everywhere that plants can grow; they have colonized the silts and sands of both fresh and salt water, a habitat which was not open to the holdfasts of the large seaweeds, and from there they grow to the snow line on the mountains and to the edge of the coldest, saltiest and driest deserts. It must, however, be remembered that a careful look in most of the habitats dominated by flowering plants will reveal algae growing in their shade, and to date the angiosperms have not challenged their floating home on the high seas. However, there is no getting away from the fact that on dry land they are the tops, so let us take a look at them and find some reasons.

Leaves

These have all the features necessary to be efficient organs of photosynthesis – a protective epidermis, well-laced with stomata; a system of air spaces through which carbon dioxide and oxygen can diffuse right up to the chlorophyll-bearing cells; and a transport (vascular) system which can transport raw materials (water and mineral salts) via the xylem, and products (sugars and chemical messengers) via the phloem. This is equally true of the leaves of the gymnosperms and ferns. However, the leaves of flowering plants have the greatest range of size, shape and hairiness.

Transportation in the Stem and Root

The xylem of the flowering plants consists of a mixture of tracheides and vessels. The latter, which offer less resistance to the passage of water, are rarely found in the other land plants.

The phloem of flowering plants is unique in that each sieve tube is accompanied by a companion cell, both of which were formed from a single cell. The companion cell only contains a nucleus and functions in part to control the activity of the sieve tube. Exactly how far this control regulates the transport of sugars is not known, but the sieve tubes can be completely closed off during periods of adverse weather, a fact which coupled with the deciduous habit could certainly fit them for life in seasonal climes.

Reproduction

Angiosperm means 'enclosed seed' and it is in the complete enclosure of the reproductive apparatus within the ovary, and the seed within a fruit, that we see the ultimate double-wrap protection.

This enclosure has led to the possibilities of many mechanisms of pollen transport

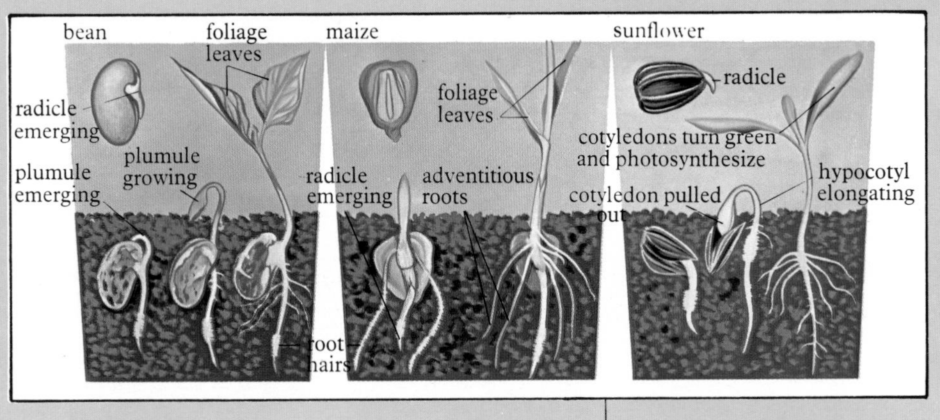

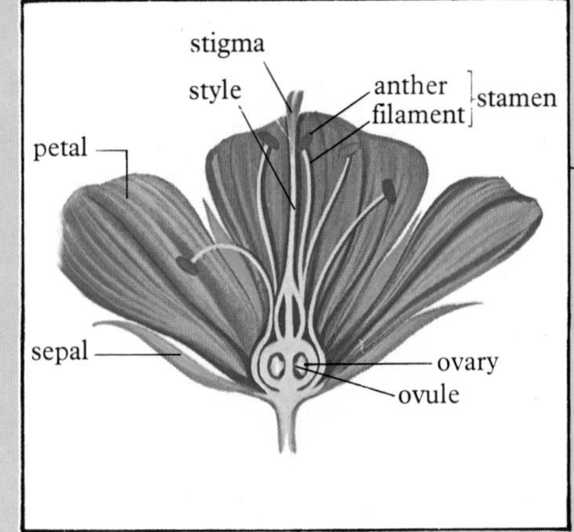

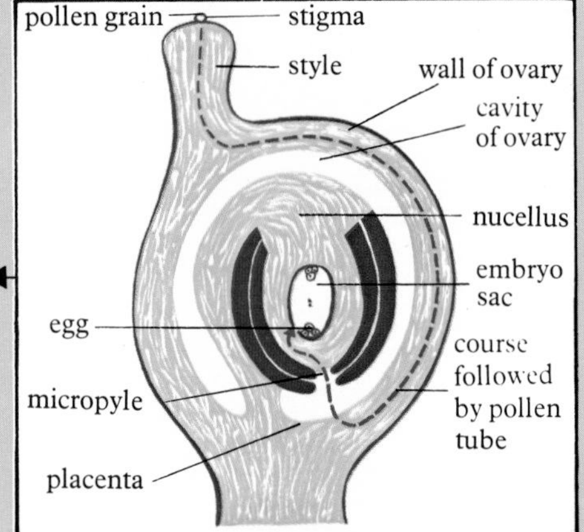

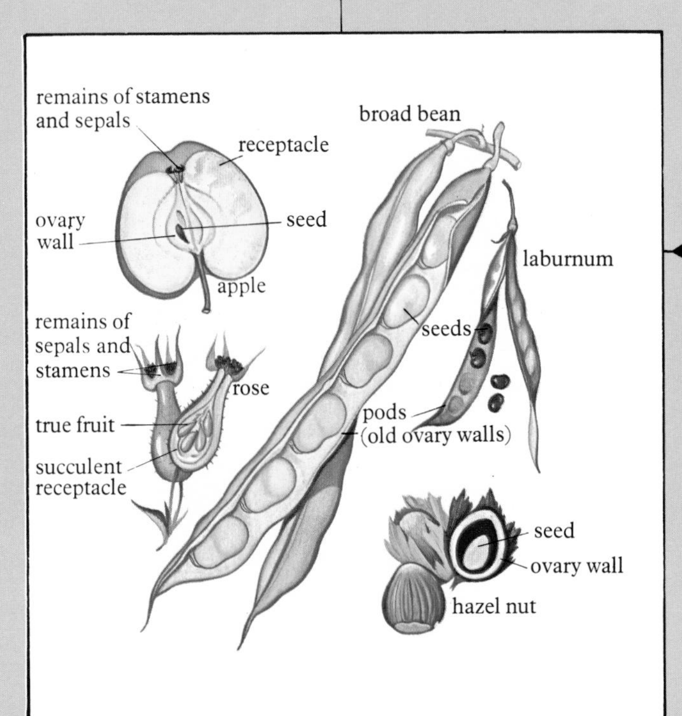

other than wind and water – insect, bat, bird – and myriad adaptations of shape, colour and scent of the flower for attracting and guiding the pollinator. In this way the flow of genetic information between the two parents is canalized, for it is no longer dependent on haphazard wind dispersal. Such canalization must greatly speed up the process of speciation. Adaptations of the fruit have made possible a similar range of dispersal agents for the seed.

Another effect of enclosure has been the need for a receptive stigma and a style through which the pollen tube grows to the ovule. What better place to erect barriers to ensure that only the right pollen grain gets there – an incompatibility mechanism exactly where it is needed, again speeding speciation.

Double fertilization is another unique feature of the flower bearers. The pollen tube growing down towards the ovary contains three packets of genetic information, a tube nucleus and two male nuclei (all that is left of the sperm). When it finally reaches its destination the tip of the tube ruptures and the tube nucleus disintegrates. One male nucleus enters the gate formed by the synergids and fuses with the ovum to form a zygote, the first cell of the next generation. The second male nucleus enters and fuses with the two polar nuclei in the middle. This triple-fusion nucleus divides to form the endosperm which will nourish the zygote as it develops into a seed.

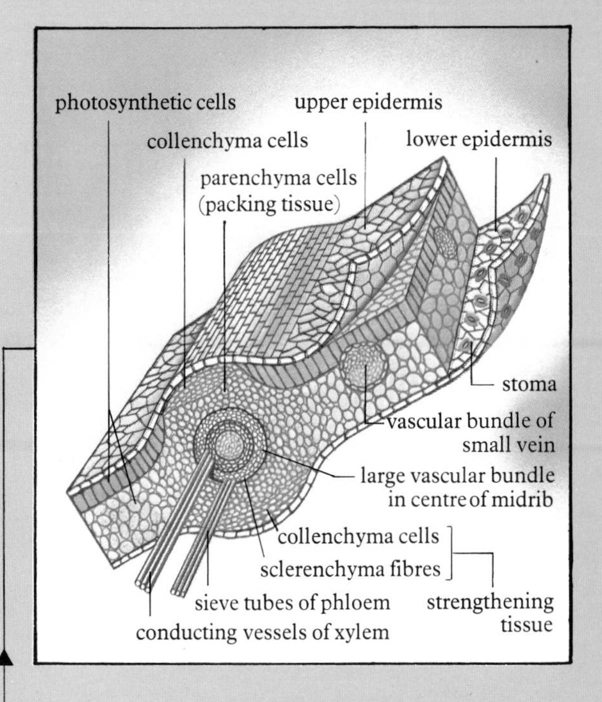

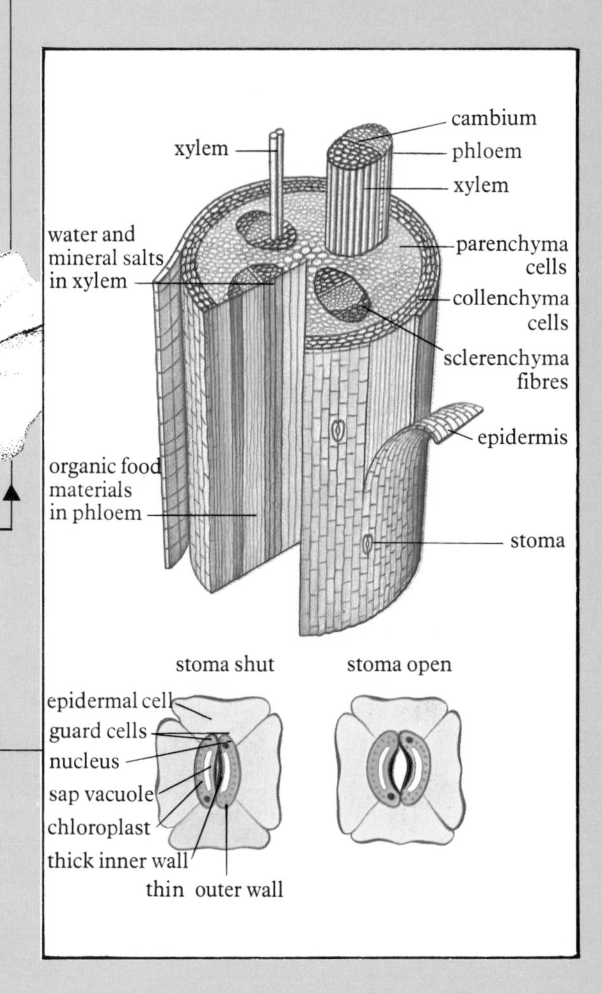

This may sound complicated, but think of the advantages. The embryo develops, nurtured by a tissue which has genetic information from both mum and dad. Most important of all, the act of double fertilization not only produces the zygote but it also triggers off the formation of the endosperm which requires a lot of energy. In the gymnosperms the food stores are laid down and a seed is produced, even if fertilization is a failure – what a waste. In this respect the angiosperms can be compared to the placental mammals in which no extra food is supplied to the developing embryo until it has been implanted. They can thus be said to be the most advanced group in the plant kingdom. No wonder then that they rose to dominance.

It must, however, be remembered that without the decomposers doing their jobs in the vegetation of the world they would not flower for many years. Success is in the eye of the beholder and I must agree that a world without flowering plants would be a much less beautiful place.

the organism produced and perfected by the process of natural selection, and the fitness of the environment itself. These two are irrevocably interlocked and interrelated. If the commonest most soluble element in the earth's crust had been arsenic then life would have either not happened or it would have happened by making use of that element. If the temperature of the earth had been much higher then covalent carbon chains would not have been viable and perhaps silicon would have played the backbone role to life. Continental drift certainly played a major role in shaping the land masses of the earth making them fitter places and, thus, increasing their potential to the evolution of the land plants and animals.

For 160 million years the cone-bearing plants, cycads and conifers, held sway and it was within the confines of their great forests that the giant reptiles evolved, held sway and had their day.

The majority of the dinosaurs had socketed teeth, and were in that obvious way more advanced than the Tuatara. During their term of office the dinosaurs, both large and small, expanded to fill all the niches that existed within the flourishing vegetation, herbivore, carnivore, omnivore. It was from their ancestral stock of socket-toothed reptiles, the thecodonts, that the first flying vertebrates like the pterodactyl arose. The plants were there providing the potential energy and evolution again responded in her now time-honoured way, using that potential by 'explosive' adaptation of the animal stocks that were then in existence.

What was it then that brought about the end of the rule of the reptiles, and the giant pteridophytes of the Carboniferous? Was it simply replacement by an organism that was better fitted to compete the others out of existence? Or was it due to some massive change in the environment which rendered the large plants and animals less fit to play a demanding role in the new patterns of potential?

Turning to the first phase of adaptive radiation way back in the Cambrian period, most of the great phyla of animals that made their appearance at that time are still with us. Granted some have been greatly modified for survival but few of the great stocks have actually disappeared. The simple fact that the environments of the sea are much more stable than those of the land and have been throughout evolutionary time could account for this. The environments of the sea 'nurtured' evolution on a grand scale and they have gone on 'nurturing' its products ever since. Changes, like the disappearance of the trilobites and graptolites, are best explained by their place being taken by more advanced forms, probably against the background of some environmental change.

The demise of the coal-swamp flora is probably best explained as an effect of major climatic change. The evidence we have to go on clearly shows that for much of the period of their rise to dominance, the floras of the whole world were very similar. This indicates that either the plants themselves were very amenable to a whole range of conditions or that the climates across Pangaea were very similar. However, by the end of the Carboniferous we find that marked differences were becoming apparent. A central belt of land, which now constitutes parts of Western Europe, North America, South America and south-east Asia, was still covered with lush tropical swamp forests. The regions to the north of this belt (the area that now makes up north Asia and north-east Russia) and to the south (the area which now makes up the land masses of the southern hemisphere) were covered with forests of a very different type. Many of the members of these showed the signs of growing in a temperate climate with alternating periods of rapid and slow growth, a fact recorded in the annual rings of their woody trunks. The southern forests of Gondwanaland were much poorer, both in the range of different plants and in the variety of insects, than their northern counterparts. A fact that could easily account for the latter difference is that the southern hemisphere was just entering on a phase of glaciation.

Whether this was due to the movements of the continents which were then taking place or some other factor, we do not know. However it is safe to say that marked climatic changes were on the way and that in all probability the giant pteridophytes were not adaptable enough to keep up with the changes.

It is great fun to speculate on reasons, especially for events that happened so long in the past, and there is the possibility of an interesting link up between the massive plants, the coal deposits and the changing climate. When we burn coal to release the sun's energy that was locked up in it some 300 million years ago, we are also releasing equally ancient carbon dioxide. During the Carboniferous period all that carbon was removed from the natural environment and was locked up out of oxygen's way. If part of that carbon came from the atmosphere, and there is no reason to suspect that it did not, and if it was not as rapidly replaced, the meagre stocks of carbon dioxide in the atmospheric envelope would have dropped. This would have had far reaching effects for, if such a drop were maintained over a long period of time, not only would overall photosynthesis slow down but the surface of the earth would get cooler. The atmospheric envelope, and especially the carbon dioxide in it, absorbs infra-red (heat) energy from the rays of the sun and thus acts as a gigantic greenhouse. Any reduction in the amount of carbon dioxide present would reduce the greenhouse effect and that could spell ice-age trouble. It all fits, but so do so many other groups of facts. It is fun to speculate, but only further research will recall the whole truth.

Returning to the demise of the dinosaurs, it is much easier to invoke climatic change brought about, at least in part, by the movement of the continents as the reason. How else can one explain the presence of coal deposits that were laid down in tropical swamps way down in the cold Antarctic? Also, the end of the dinosaurs is contemporary with the most recent phase of continental drift. In such a changing environment adaptability would be a main key to success and the reptiles were all faced by one severe limitation–they could do very little to control the temperature of their bodies, and were hence at the mercy of the environment in which they lived.

Just as the plant kingdom could

not emerge from the sea until it was able to control its own hydrature (amount of internal water), so too were the land animals restricted to the warmer regions of the earth until they could control their own temperature. Thus, we find poikilohydric plants which can only grow in free water or immediately after rain, including the algae, mosses, liverworts and the gametophyte generation of the ferns and their allies; and homoiohydric plants that can live in all but the most extreme dry deserts, including the cycads, conifers, flowering plants and the sporophyte generation of the ferns and their like. Similarly, the amphibians and reptiles are poikilothermic (cold-blooded) and they live mainly in the tropics or undergo long periods of hibernation, and the mammals and birds are homoiothermic (warm-blooded) and they can roam the climates of the world.

The plant kingdom was thus 'dragged' out of the water as it answered the challenge of the poten-

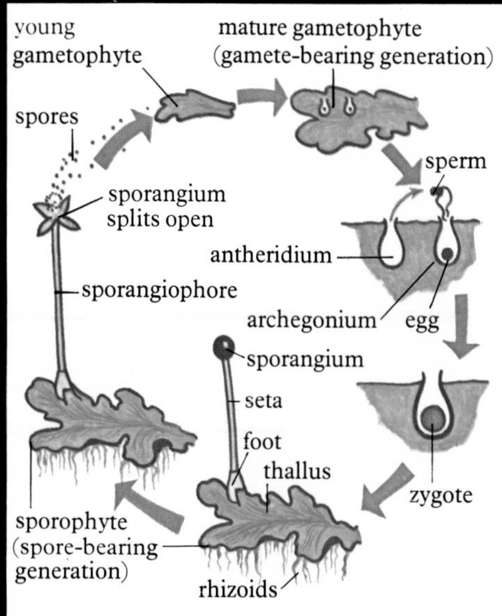

The liverworts

The mosses, liverworts and hornworts, although small in stature, include about 25 000 contemporary species. They are poikilohydric, always growing in or close to water or having the ability to aestivate (shut up shop) during periods of drought. The two generations of their life cycle are more or less of equal stature although in the majority of cases the gamete-bearing generation is the most conspicuous and lasts the longest. On the liverwort shown, antheridia and archegonia are both on the same plant. In some liverworts, however, the male and female organs are borne on separate plants. The sperms from the antheridia swim to the egg of the archegonium. After fertilization, the zygote swells to form a 'foot', embedded in the gametophyte tissue. A stalk, called the seta, arises capped with the sporangium. When ripe the sporangium splits open – usually by the unequally thickened cell walls drying and splitting – releasing the spores into the air to germinate into new gametophytes.

Apart from their shape and mode of reproduction, the most important feature which separates the gametophyte from the sporophyte is that the epidermis of the latter may have real stomata – shades of things to come.

The plant kingdom and its major firsts

It is very satisfying to draw phylogenetic (family) trees linking all the various groups of plants together and suggesting what evolved from what. It can also be very misleading, especially if you have not got all the evidence to hand. As we are far from that happy state of affairs all the diagram attempts to do is show the range of evolutionary endeavour and the major breakthroughs made by the plant kingdom.

prokaryotes

eukaryotes

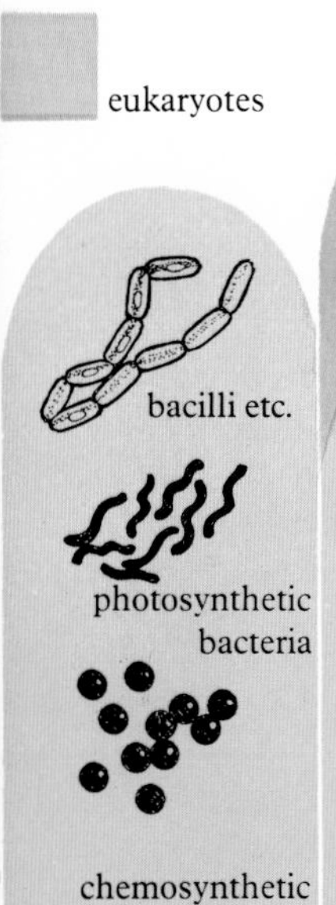

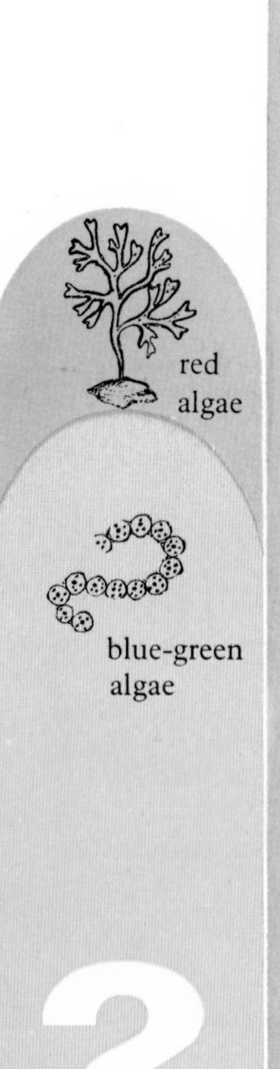

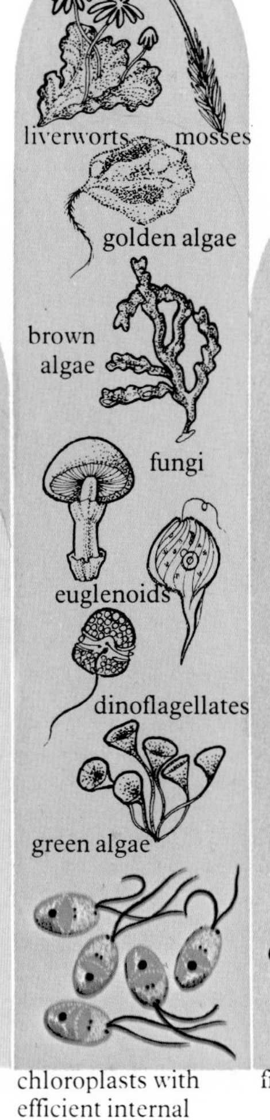

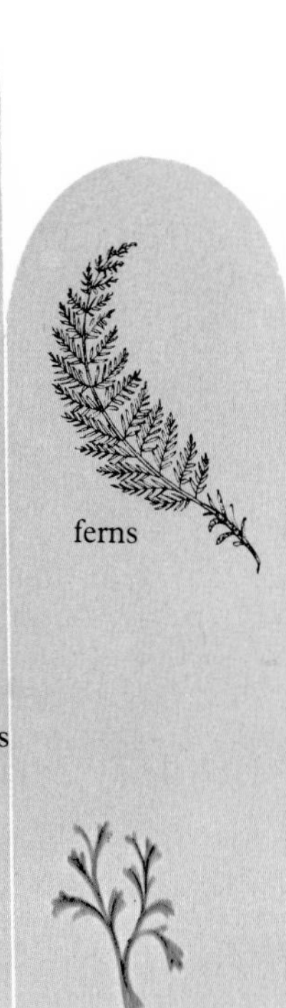

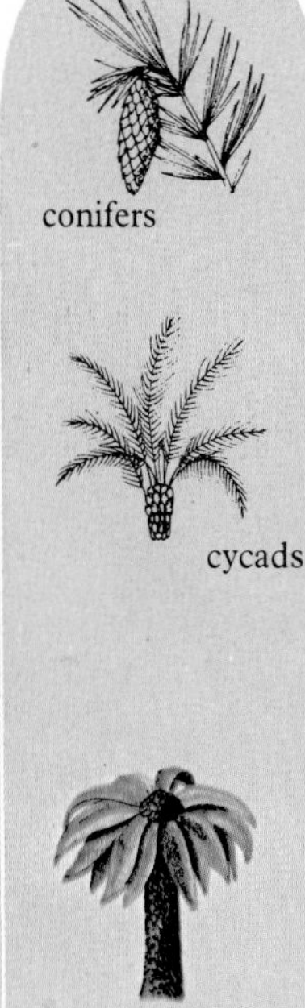

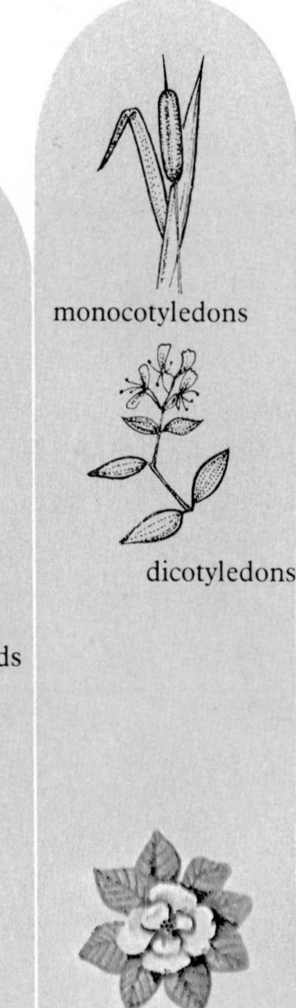

organization within a membrane | first oxygen producers | chloroplasts with efficient internal organization | first stems | first leaves | first seeds | first flowers

poikilohydrous | **amphibious** (split personality) | **homoihydrous** land plants

Have form will function

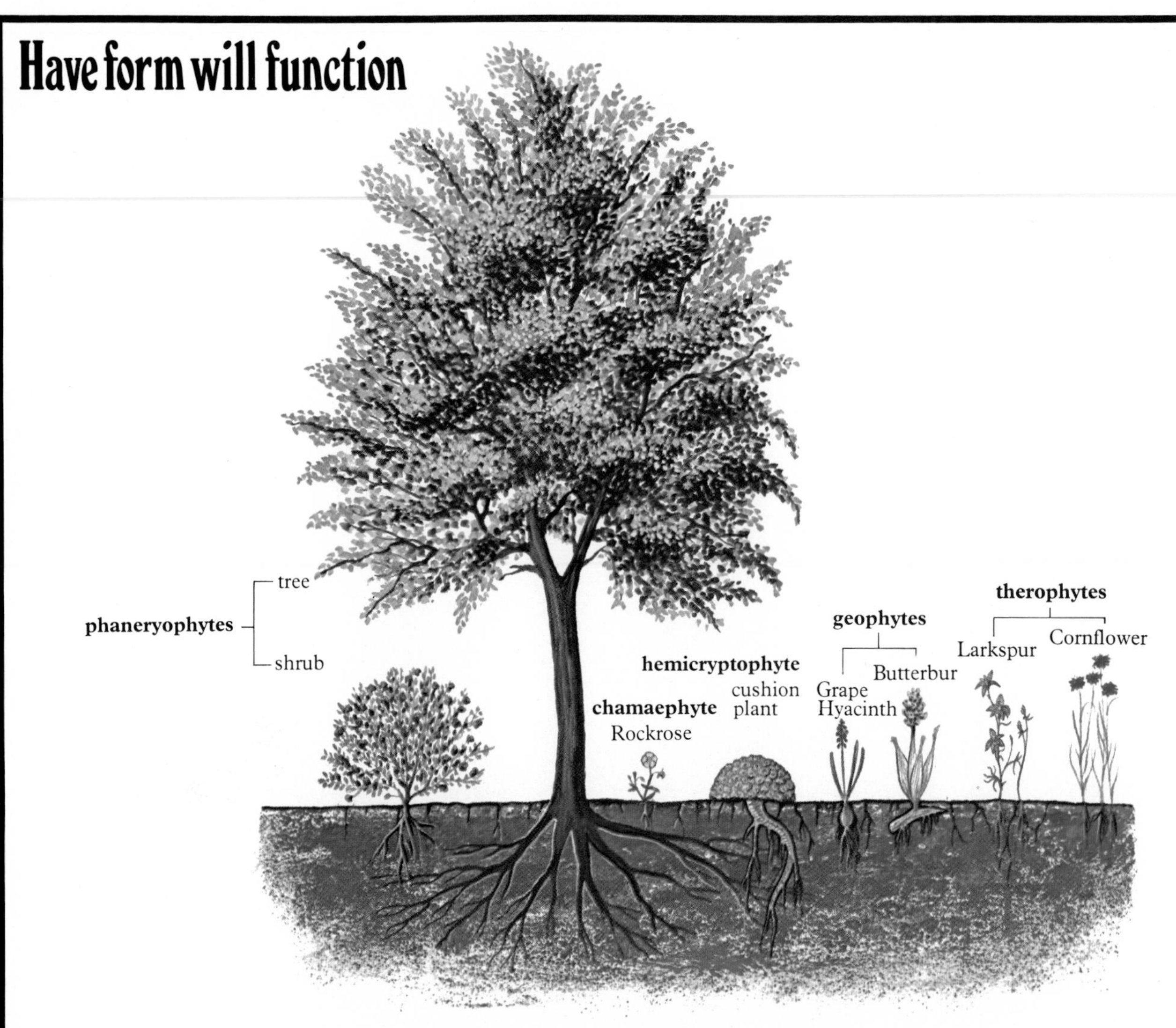

According to a botanist by the name of Takhtajan there may be up to 250 000 species of flowering plant arranged in some 304 families. A Danish botanist, Raunkiaer, grouped them, using a system which, for the land plants, depends on their form during the most unfavourable season (in a temperate climate, winter).

1 *Phanerophytes* woody plants with buds more than 25 centimetres (10 inches) above soil level.
2 *Chamaephytes* woody or herbaceeous plants with buds above the soil level but below 25 centimetres (10 inches).
3 *Hemicryptophytes* herbs (rarely woody) with buds at soil level.
4 *Geophytes* herbs with buds below the surface of the soil.
5 *Therophytes* plants which pass any unfavourable season as seeds.

Raunkiaer's life forms are somewhat akin to Gams' functional groups (see p8), but they are of wider application in that the distribution of some of them shows a distinct correlation with climate and habitat. For example, phanerophytes are never found in the coldest climates; hemicryptophytes are common in cold habitats when snow protects the buds through winter and they are also common in vegetation that is subject to intensive grazing; geophytes and therophytes find their optimum habitats in warm regions which are subject to a long dry period and/or to fire.

Preceding pages The bulk of modern conifers have needle-like leaves with sunken stomata, both of which mean that they need lose only very little water. This is ideal for an evergreen plant living in cold climates where the water supply in the soil may be frozen solid for much of the winter. However, even if they lose no water at all the leaves will have to tolerate the icy blast of the wind and temperatures way down below zero. Certain conifers have the toughest, most frost-tolerant leaves in the world. Nevertheless a nice blanket of snow must help to keep them warm (0°C (32°F) must be a lot better than –30°C (–22°F)).

tial of the estuaries. The animal kingdom was similarly faced with new challenges, new opportunities in a changing world, which they faced with a warm heart.

Again it is fun to seek out interrelationships. The demise of the giant reptiles went hand in hand with the flowering (if you will excuse the pun), of the angiosperms. Apart from their two important vessels, they are in part set aside by their ability to produce a range of chemicals called alkaloids.

Although the alkaloids are nitrogenous organic substances, and thus come within the compass of the term life chemicals, they really deserve the name of the death chemicals for they include such potential poisons as atropine, cocaine, morphine, quinine and strychnine. All of these are poisonous to higher animals unless they are used in moderation.

Not all flowering plants produce alkaloids in sufficient quantity to make them dangerous but they are widespread constituents of the tissues of a number of plant families. Now one thing that the giant herbivorous dinosaurs could not do in moderation was eat, and several hundred kilograms of new vegetation well-laced

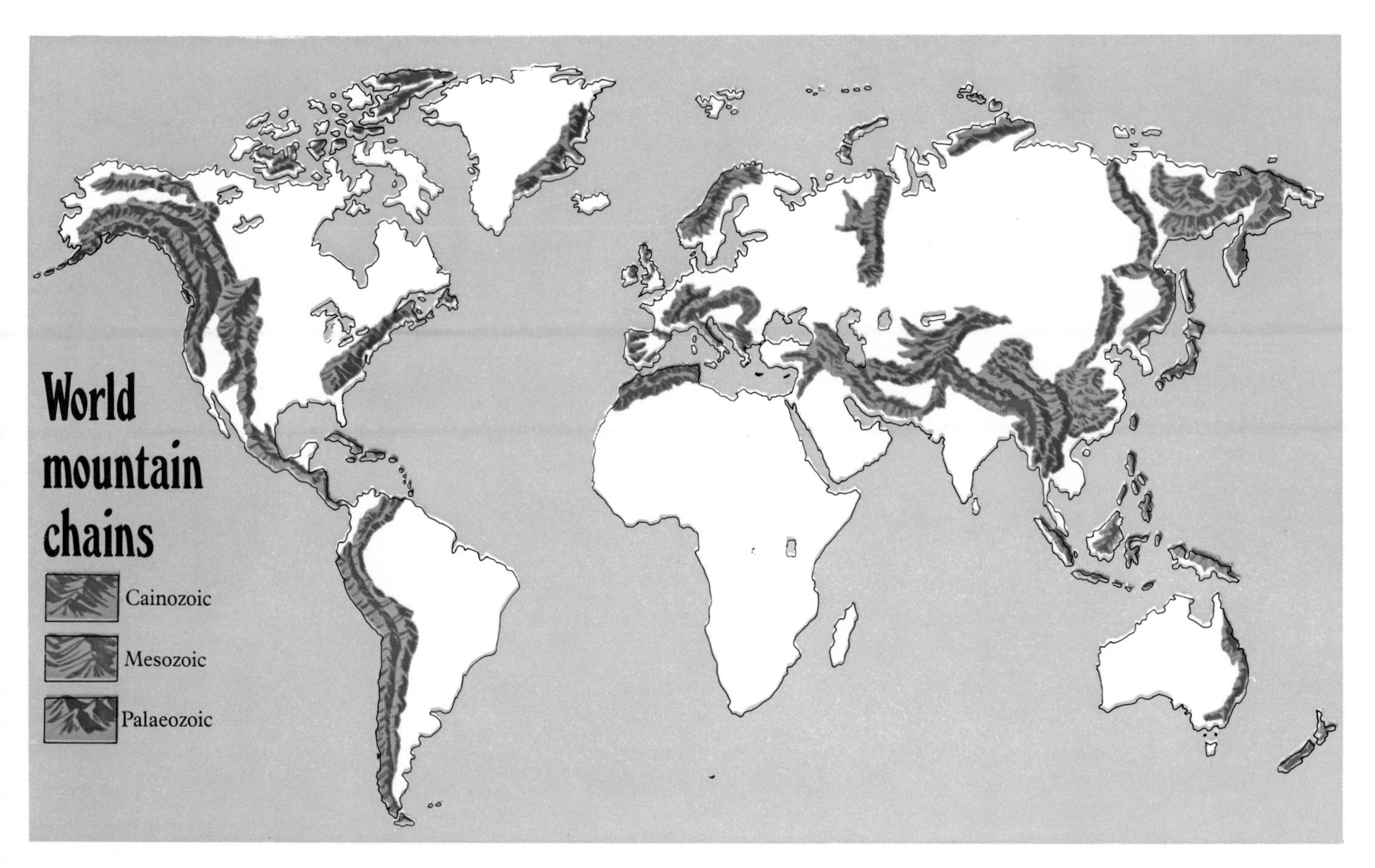

Mountains are built not created and most of them have been built in comparatively recent times. In fact, many of them are rising beneath our feet whether we are on skis or not.

with alkaloids could have been sufficient to hurry their extinction. Again, this is pure speculation but that is the stuff of hypothesis on which all science has its foundations. After all, not that long ago the concept of continental drift was thought to be preposterous.

As the continents eased apart, the flowering plants and the mammals began to take over those environments in which their own special adaptations gave them some real advantage. The reptiles found themselves niches in the warm, and especially in the drier, parts of the world. Likewise, the gymnosperms went their own way, the cycads to a restricted distribution in the tropics, while the conifers exploited their tolerance of the colder polar and alpine climates. Part of their success lay in a rapid rate of growth correlated with their being evergreen. The majority of conifers have needle-like leaves that not only lose very little water but become frost-hardened, so they are not damaged by the very cold conditions of the long polar winters. Their leaves are, therefore, on the plant ready to start photosynthesis as soon as the climate improves in the spring.

The flowering plants, on the other hand, developed the ability to be deciduous, that is to throw their leaves, and thus most of their winter problems, away at the onset of adverse climatic conditions. Among their contemporary representatives we find the tallest trees which have ever existed; shrubs and herbs, fitted to play lesser roles in the structured forest systems; and geophytes with bulbs, corms and rhizomes, underground organs which store sufficient food to tide them over periods that are too dry or too cold for normal growth. Last and very often least, especially in stature, are the annuals, plants which can pack their complete life cycle into a year or often much less, seed to seedling to maturity, flower, fruit and back to seed once again–the ultimate way of fitting into a changeable climate.

The changes were certainly coming in thick and fast, for not only were the continents themselves moving through different longitudes, but as they went they created the stresses and strains in the earth's crust which threw up the greatest of the mountain ranges. Massifs like the Alps and later the Himalayas began to act as barriers to the flow of the main air masses, setting up the pattern of local weather and future climatic change.

Perhaps most important of all in the changing arena of evolution were the interactions of the plants and animals themselves.

Around 60 million years ago, hooved animals made their appearance–the odd-toed rhinos, horses and tapirs and the even-toed, cloven-hoofed hippos, pigs, deer and cattle, all of which are browsers or grazers. The flowering plant responded with plants which grew closely adpressed to the ground, plants which tasted unpleasant and/or were poisonous, and plants with leaves that go on growing after their tops have been chopped off. Most famous of the latter are the grasses which evolved long before man came on the scene with his lawn mower. Which came first, the adaptations or the stress, is difficult to say. The result is, however, much the same because once the modifications were there, the

The Cainozoic era

The first mammals made their appearance in the Mesozoic era, the end of which was marked by the demise of the ruling reptiles. The new era of the Cainozoic opened up with the bang of another 30 million year evolutionary explosion (25 to be exact, the length of the Palaeocene and Eocene epochs) which produced all the modern groups of mammals – the pouch-bearing marsupials and the live-bearing placentals, including the prototype whales, bats and grass-eaters. Like the flowering plants their key to success was to protect the next generation until it was ready to stand up by itself to the rigours of life on the land. With little or no competition from any other large land animals, the changeable environments of the dry earth were their arena for success.

plants that had them would be at an enormous advantage compared to those without. So it has been throughout evolutionary time, potential builds on potential.

I was awoken from my evolutionary daydreams by Otto Gram starting up the engine of the helicopter. It was time to go, but before we headed for home we had one more visit to make, the active peak of Ngauruhoe itself. We began making our way back and forth over the black cinder cone until we hovered looking down into a boiling cauldron of dilute sulphuric acid. The surface of the crater itself was hidden by an immense uprush of steam. Otto shouted above all the noise, and yet in tones of great respect, 'that's a very angry mountain'.

If we needed any reminder of the enormous natural forces associated with the process of continental drift, it was there below us. In the same way if we need proof of the effect of continental drift on the evolution of the flora and fauna, New Zealand provides it in abundance.

On all the evidence to hand it does seem safe to conclude that this small mass of land split away before the main expansive phase of evolution of either the mammals or the flowering plants. There is also no evidence to suggest that its land surface ever felt the tread of the great dinosaurs or, indeed, any grazing quadruped. It would appear that the only large animals that have ever had any massive effect, grazing on and trampling its vegetation, were the birds which arrived quite recently.

When Polynesian man first arrived on the New Zealand scene he found a land that was still in the late stages of emerging from the effects of the last glaciation. The bulk of the lowlands were covered with extensive forests, and in the less favourable climates with grassland. The forests, of which there are some remnants more or less in their natural state still left, were of great interest to the first naturalists who came more than 800 years later. There were forests of tree ferns, from the trunks of which hung

Although the southern beeches (*Nothofagus*), resemble their northern-hemisphere counterparts in many ways, they do have one main difference – they are evergreen. However, it must be remembered that the northern beeches (*Fagus*) often appear very unwilling to shed their last year's leaves, especially when they are growing in sheltered positions. No tree actually throws its leaves off; it just loosens them and then the wind blows them away.

Tmesipteris the only surviving relative of *Psilotum*. If the plants of the coal swamps could have survived anywhere in the modern world, this was the most likely place. However, the ice ages probably put paid to that. Perhaps most famous of all New Zealand's plants is the Kauri Pine (*Agathis australis*) although it really should not be called a pine as it is related to the monkey puzzles. It is cone bearing and ranks among the mightiest trees in the world, growing up to 70 metres (230 feet) high with the first 30 metres (100 feet) of the trunk unbroken by any branching. It is a magnificent timber tree, a fact that has led to the destruction of all but the most inaccessible forests. This is very unfortunate because it takes 1 500 years for one of these big trees to grow to maturity. Other broadleaf conifers are the podocarps, which form extensive forests in places and are also found mixed with hardwoods. The most famous hardwoods of New Zealand are the southern beeches, *Nothofagus*, which fill a similar role to the beeches, *Fagus*, of the Northern Hemisphere.

One very interesting fact is that 65 per cent of the native trees of New Zealand are completely dependent on birds for their dispersal. The fruits of the trees are eaten by birds, the tough seed passing through their digestive system to emerge at the other end ready to germinate in their very own pile of manure.

Apart from their sweet tooth for the fruits of trees the birds of New Zealand are themselves remarkable, particularly for the high incidence of flightless species among their numbers. The latter fact can best be explained by invoking the existence of a terrain devoid of predators or competitors. In the absence of predators the importance of flight would have been reduced. Why bother to fly if you are a herbivore with no competitors on the ground or, perhaps more appropriately, in the field? Thus it was that certain flightless birds evolved to fill several niches which over much of the world are the strict preserves of the mammals.

Most spectacular of these were the moas, the largest being *Dinornis maximus*. This giant among birds ate leaves, twigs and coarse grasses, a fact that marks it out as king of the 'lawn moas'. In the absence of teeth it employed the grinding action of 4 kilograms (8·8 pounds) of gizzard stones to prepare the tough plant material for digestion. A 4-metre (13-foot) herbivore, only surpassed today by the giraffe and the elephant, this great bird was doomed to extinction along with the other twenty-four or so lesser members of the moa clan.

Whether it was New Zealand's changeable climate or solely the arrival of man that brought to an end the reign of these herbivorous birds, is not clear. The Maoris constantly hunted them and their introduction of dogs and pigs could not have aided the survival of the flightless birds. The abundant remains of moas in swamps and caves allow us to build up a very good picture of their way of life and there is the remote and exciting possibility of forest-dwelling species being found alive in the remote fjordland of the south-west.

The latter possibility has recently been given a boost by the discovery

of the Takahe (*Porphyrio mantelli*), a large flightless rail which was thought to have become extinct at the end of the last century. It was discovered again by an amateur ornithologist in a remote valley near Lake Te Anau in the south-west of South Island in November 1948. The whole area around this valley is now a highly controlled reserve and several birds are held in a breeding colony.

The Takahe is a remarkable bird, about the size of a large hen. It has blue, blue-green and white plumage with a massive red bill and legs to top and bottom the lot. It is a voracious grazer and appears to eat more or less continuously, mowing the herbage in front of it and laying a trail of organic fertilizer behind. Like a number of herbivores its digestion is not very efficient and an adult can produce an astounding 8 metres (26 feet) of droppings per day. It makes you wonder what *Dinornis* produced!

New Zealand did not, therefore, present a complete paradise for the evolution of the plant kingdom. However, the living treasures of New Zealand, both plant and animal, and especially the living systems of vegetation, do provide us with a glimpse of what things were like over much of the world in past ages.

It would almost be sacrilege to write about natural New Zealand without mentioning its most famous flightless bird, or rather birds, because these are three different sorts of kiwi, *Apteryx australis*, *A. owenii* and *A. haastii*. There are three species in one genus which is the only member of the family Apteryeidae, which is the only member of the order Apterygiformes–an order all of its own; you can not get much more special than that. One of the unique features of these birds is the position of their nostrils, which are at the top of a long flexible beak, ideal for poking into things and smelling out the worms and grubs on which they feed.

The kiwi holds the world record for size of egg in relation to the size of the bird. Mother kiwi lays an egg that weighs a quarter of her own weight and then passes it on to dad who looks after it for between 7 and 10 weeks. The baby kiwi emerges fully fledged and fighting fit, soon ready to defend itself with its strong legs that evolved to dig up its food. It is for this reason that of all New Zealand's flightless birds the kiwi has suffered perhaps least at the tooth of introduced mammals and now appears to be making a comeback.

The enormous problems of introduced plants and animals brought to New Zealand by the settlers were made clear to me as we flew back down the mountain across the Tongariro National Park. The lower slopes were covered with a sea of heather, and we landed in a scene that was 100 per cent Scotland. Otto leapt out to remove an alien thistle. 'Damn things,' he shouted, 'they get everywhere. Unfortunately the heather is here to stay but we try to keep the others out.'

Here was a man, himself a true pioneer, who through his own skills, hard won through long apprenticeship, had become master of his chosen environment. Yet he has time and feeling for the natural inheritance of the Lands of the Long White Cloud. However, unlike the native kiwi, Otto Gramm had learned to fly! He had to, because that is how he makes his living. Wherever there is potential. . . .

Left Three Kauri trunks and a tree fern. When it comes to volume the Kauris probably hold the world record, because their trunks do not taper but form a solid shaft of timber which may be unbranched for over 60 metres (200 feet). Unfortunately, this makes them ideal for chopping down. New Zealand also boasts the tallest ferns in the world, up to 20 metres (66 feet) in height. Here is one growing in the shade of the old Kauri trees.

Right A reconstruction of *Dinornis maximus*, the Giant Moa. If only he had lived to tell his own tale.

Below *Porphyrio mantelli*, the Takahe. Does he hold a world record? Once the Takahe was widespread throughout New Zealand, but as none had been seen during the early part of the twentieth century, it was thought to be extinct. However, in 1948 an amateur ornithologist found a thriving colony in a little-explored valley in the mountain on the western side of Lake Te Anau. The whole area is now a reserve. Breeding birds are held in captivity and appear to be doing well.

The rise of the land plants

The story of the rise of the land plants postdates the acquisition of stomata by the spore-bearing generation. There can be little doubt that the presence of a 'waterproof' epidermis, which included stomata within its make-up, allowed the shoots of the sporophyte to be an efficient photosynthesizer under a range of environmental conditions. Hence it could gain dominance over the gamete-bearing generation which could do little or nothing to control water loss from its tissues and so had to shut up photosynthetic shop when the growing got too dry.

Two great groups of plants, the Cycadofilicales (seed ferns) and the Cordaitales certainly led the way on to dry land and they paid the price of extinction before the dawn of the Mesozoic era, which is rightly called 'the age of the gymnosperms'. Evidence indicates that the former gave rise to the Bennetitales (fossil cycads) and the Cycadales (modern cycads) and the latter to the Ginkgoales (ginkgos) and Coniferales (conifers). The exact point at which the Angiospermae (flowering plants) split off, probably from the seed fern stock, is a matter of debate. However, by 100 million years ago some fifty of their families were in evidence and by the end of the Cretaceous period all of the main subdivisions of the flowering plants were in existence. The potential was there and the flowering plants radiated outwards and upwards to use it. The success of the angiosperms is manifested by the fact that today there are more than 230 000 species making the earth such a beautiful place.

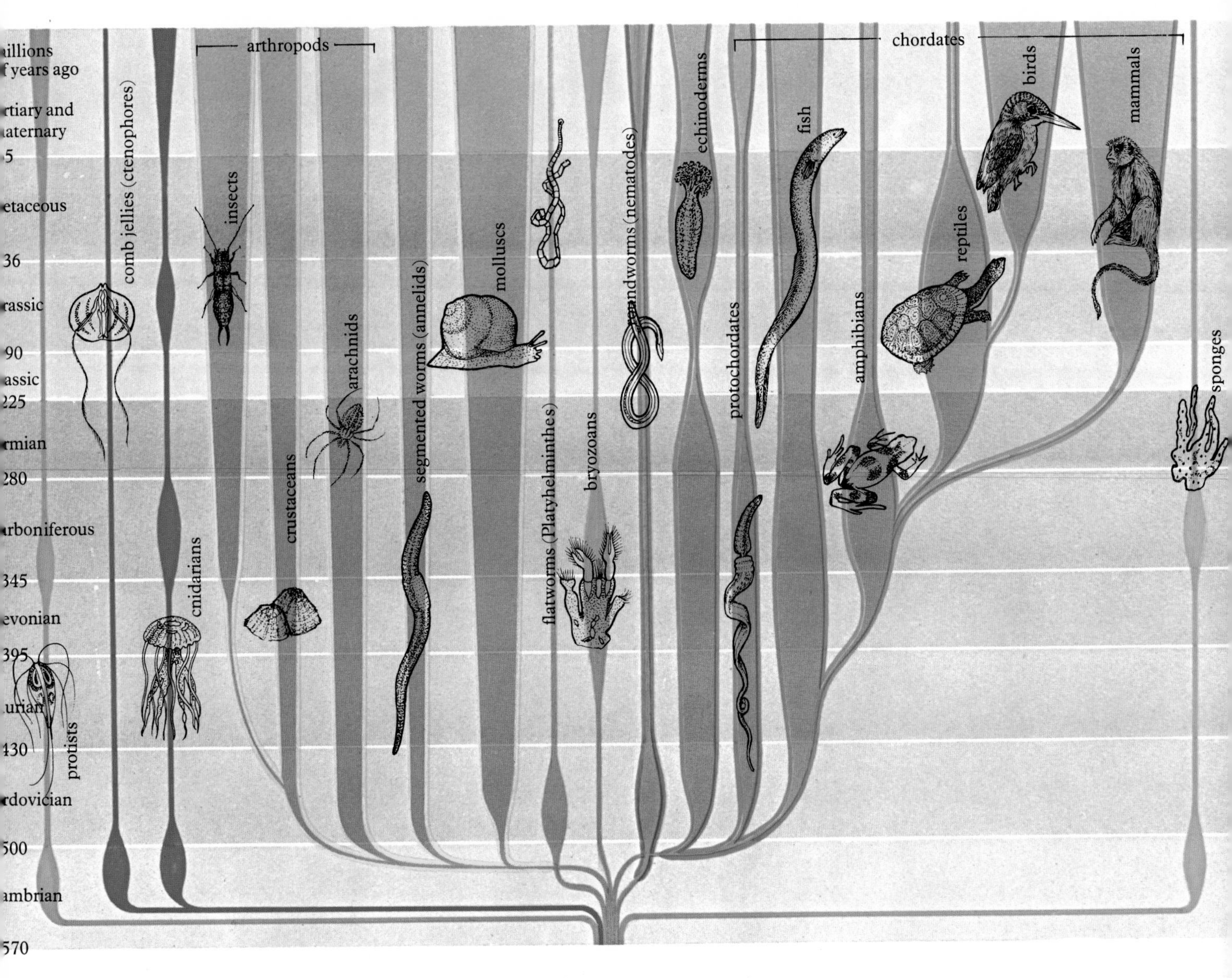

The kingdom of animals

Some of the twenty-eight phyla of animals are shown in the diagram. It would be misleading to say that the ones (and their sub-groups) selected are the most successful or the most important, for each animal group is successful in its own right and all are important in the overall pattern of life. Suffice it to say that they do represent a cross-section of evolutionary endeavour. The width of the lines shows the periods of their greatest diversity.

The coelomates are arranged radiating into the two main lines of advance, which led to the two most successful phyla of land animals, the Arthropoda and the Chordata.

Colour key

purple—radial symmetry
pink—bilateral symmetry
green—false coelom
yellow—coelom formed as new split
blue—coelom formed from primitive gut cavity

Chapter 6

Latitude zero

What better place to take a look at the products of evolution on the land than the equator in Ecuador. Here it is possible to walk from the lush tropical rain forest of the lowland up to the ice deserts of the high Andes, without stepping off latitude zero.

Such a journey not only lets you see a fantastic range of the products of evolution, but emphasizes the fact that the full potential of each environment can not be exploited by a single organism. Evolutionary fitness always relates to the role an organism plays within a complex community, and it is the community alone that can reflect the opportunities on offer. The same is true for man.

If you could join the elite band of astronauts, sit back in deep space and view the planet earth, you would have ample proof of the success of the land plants. Between the clouds of life-giving water, it would be possible to see that the land masses are marked out by a series of more or less distinct zones. The more sophisticated your method of looking, or rather sensing–infra-red, ultraviolet, computer enhancement, in fact the whole technology of the spaceman's look-book–the more detail would become evident. The zones and the patterns would in the main be types of vegetation, and in no case would you be seeing a single super plant living it up, king in its own particular environment. You would be viewing super systems of plants and animals, each one having evolved to exploit the potential of the environment of that particular area.

The early natural historians who travelled the world on foot and horseback first noted the existence of the zones of vegetation and explained them in terms of climate. One of these early travellers was Alexander von Humboldt and he was perhaps luckiest of all, for his journeys took him to the Pacific coast of South America which, is, without doubt, the perfect place in which to gain an understanding of the interrelationship between vegetation and climate. In places it is possible, in a distance of less than 200 kilometres (124 miles), to pass from the heat of the tropics up to the permanent ice of the high Andean peaks. In Ecuador you can make the whole transition and never even step off latitude zero.

The environment of the equator has a number of distinct advantages for life. First and foremost, each square metre of land should, aspect and cloud cover permitting, receive the maximum possible amount of energy from the sun, to be exact, 210 joules (155 foot-pounds) per second.

Secondly, the energy supply is on for about 12 hours and off for the next 12 every day of the year. There is no cycle of spring, summer, autumn and winter which characterizes the climates of higher and lower latitude. The only seasonality, if such it can be called, is due to variations in the amount of rainfall.

For 12 hours each day the bed rock

The zonation of vegetation on the Andes

Wherever there is potential it will be used – an aerial trip up from the coast of Ecuador to the young snow-capped peaks of the Andes provides the best visual proof that ecosystem evolution obeys this law to the letter. The vegetation of each altitudinal belt fits the environment like a well-tailored garment, spelling out the potential of equitable temperatures and sufficient rain in terms of stature of the dominant trees, biomass and annual productivity.

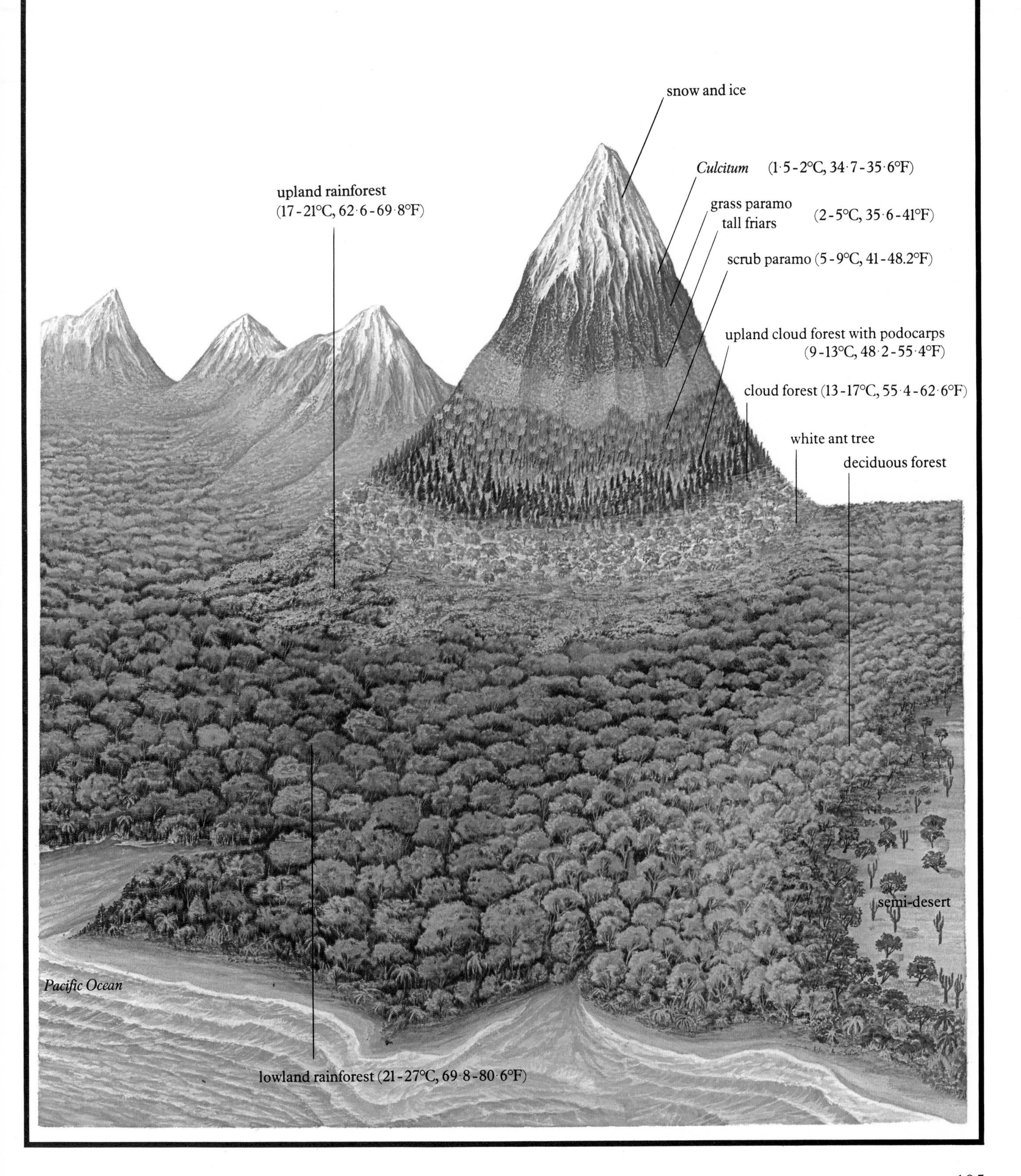

acts as a heat sink, the surface layers warming up very rapidly and for the 12 hours of night it acts as a radiator loosing heat to the air above. In the perfect 12 hours on 12 hours off (diurnal) climate, the soil below about 30 centimetres (12 inches) depth comes into equilibrium with this heat flux and its temperature remains more or less constant throughout the year. It is thus possible, using nothing more sophisticated than a spade and a thermometer, to ascertain a rough figure for the mean annual temperature of any location. The further you are from the equator the less meaningful would the mean annual soil temperature be owing to the large seasonal fluctuations of the air temperature above.

The air acts as a thermal blanket, heating up and cooling down, buffering the fluctuations of the surface temperature. At sea-level the air is densest and, therefore, has its greatest effect. Passing upwards the air gets thinner and thinner, its thermal blanket effect diminishes and the diurnal temperature fluctuations are very rapid. The mean annual temperature at sea-level on the equator in Ecuador is about 27 degrees Centigrade (80·6 degrees Fahrenheit) while at the snow line on Mount Cayambe which sits astride the equator, it is 1·5 degrees Centigrade (34·7 degrees Fahrenheit). It is possible by deviating a few degrees from latitude zero to select a series of areas with an equable rainfall climate–that is, one in which there is sufficient rainfall throughout the year to sustain plant growth and hence an evergreen vegetation. These areas must represent the best conditions possible for the maintenance of plant life.

The diagram (p 105) shows an idealized section across the wet Andes indicating the main zones of tropical evergreen vegetation that would be encountered on the way, together with the mean annual temperature range in each. Unfortunately, little detailed work has been carried out on all the zones and our knowledge of the make-up of each vegetation type, let alone the factors which contribute to the differences, is extremely rudimentary. Nevertheless, some overall meaningful observations can be made.

The two lowest zones can, for convenience, be lumped into a single vegetational unit, wet tropical lowland forest. Throughout this temperature/altitude environment there seems little or nothing to limit the growth of plants, and these forests, together with the tropical rain forests of the rest of the world, rank among the most diverse and productive living systems on earth.

Our detailed knowledge of the lowland forests of Ecuador stems from a 2-square-kilometre (0·8-square-mile) forest reserve called the Rio Palenque which is run jointly by the Universities of Miami and Quito for teaching and research. It was, in part, the brainchild of Calaway Dodson, an American botanist who specializes in studying the relationships between insects and plants. The following story, which he told me, emphasizes the importance of basic scientific research which at first sight may seem completely useless.

When they first visited the site prior to purchasing it as a reserve they noticed a large tree that had fallen in part in the river. Five years later when they came to build the field station the fallen tree was still there, its trunk intact, despite the potential ravages of termites, fungi, bacteria and all the other members of the tropical trash-can brigade. The remarkable non-rot wood was found to be running with natural oil and was used in the construction of the field station. Study of the tree by one of Dodson's colleagues, A. Gentry, showed that it was a species of *Persea* new to science, and he named it *Persea theobromifolia*. Careful search revealed a number of other specimens in the reserve area and study showed that it could be propagated with ease.

The potential of this tree to tropical forestry has yet to be researched and one can only begin to guess what other important plants are awaiting discovery in the, to date, almost unexplored forests of this and other parts of South America. Unfortunately, today the forests are being destroyed much faster than they are being studied.

Research at Rio Palenque has revealed so far the presence of 1 326 different vascular plants and the list is being added to all the time. About 850 of the plants are thought to be members of the natural flora of the forest, the rest being immigrant tropical weeds brought into the area by man, for today the reserve stands alone in a sea of oil-palm plantations and ranches. Already 322 different birds have been recorded from the area and the list of amphibians, reptiles and insects being built up is very impressive. Apart from the enormous diversity of the life of the area, the sheer size of some of the forest trees is equally impressive. It is difficult to measure the exact height of a tree, especially when it is growing on the side of a steep hill and is surrounded by other trees. There are, however, a number of forest giants present in the area which in all probability top the 70 metre (230 foot) mark and one I estimated to be almost 80 metres (262 feet) tall. It was a fig tree that would have provided decorum for a multitude of Adams, and flowering plants do not come much bigger than that.

With such an enormous diversity of life it is difficult to know what to select to talk about. There are clumps of grass (bamboo) that spring up to more than 30 metres (98 feet) in height, the younger shoots of which would bring the largest Chinese 'take-away' to its knees yet, when tapped in the right way, provide a ready supply of pure sparkling water. Diminutive hummingbirds feed from the flowers of exotic bromeliads that crowd the crotches of the largest trees. The bromeliads deserve a special mention. Members of the Pineapple family, they are among the most conspicuous of the epiphytes, and are going to be with us throughout our journey. One may see a spider which looks so like the orchid flowers among which it sits that unsuspecting insects which visit the flowers for nectar fall an easy prey. Everywhere one looks there is something to marvel at. Each interrelationship is a part of the gigantic living system, a part and a product of the story of evolution.

It is so easy to stand within the structure of the forest and forget that it is all alive, the pillars (tree trunks), walls (lianas and vines), windows (cross-hatched branches) and the arched ceiling (leaves) above are not made of inanimate bricks and mortar but are all composed of water held in a matrix of structured living chemicals, at a rough estimate around 50

Above Epiphylls on the leaf of a fan palm. Palm leaves are often very large and the tough ones last for a long time, making them an ideal perch for any mini-epiphytes. Here an old leaf, partly eroded away, is covered with many colonies of lichens (the grey patches) and some small mosses and liverworts. Already a larger epiphyte is making a bid for space on the leaf but it will not get far because the support leaf is now so shaded that its own end is very near.

Overleaf The White Ant Tree, *Cecropia albida*. While its ants are at home it stands out alone, at around 1 700 metres (5 580 feet) above sea level. This is just one of many thousands of examples of a myrmecophyte which can be seen in the forests of Ecuador. Myrmecophytes are ant plants, or rather plants which live in symbiotic union with ants. The plant provides a convenient home and the ants guard it, often with their lives, from other marauding insects.

thickets of a lesser bamboo about 3 metres (9·8 feet) tall, entwined with vines and interspersed with plants such as roses, blackberries and exotic members of the heather family, all of which make a botanist from temperate latitudes feel very much at home. There are, however, an overwhelming number of plants from purely tropical families, enough to dispel any ideas of higher latitudes.

What is perhaps most surprising of all is that many of the plants that were exclusively epiphytic members of the sub-canopy in the lowland forests now find a niche on the ground growing on the rocks, boulders and especially on the vertical cliffs.

Tree ferns abound in the damper spots, their prothalli finding ample water in which to release their swimming spermatozoids in order to complete their very unequal life cycle. It is at this level that the ferns really come into their own, especially the filmy ferns which festoon both trunks, branches and rocks. The spore-bearing generation of many of these should, like their gametophytes, be considered to be poikilohydric, for they have so little control of the loss of water from their delicate leaves that even a short period of exposure to direct sun can lead to wilting and death.

It is interesting that in a zone which abounds with so many 'amphibious' plants, there are many frogs and toads, which make their presence felt after rain by their hopping and croaking. Among their members one of the least amphibious of the amphibians may be found. It is the marsupial frog which, as its name suggests, has a pouch located on the female's back in which a brood of about twenty eggs develops emerging fully frogged to take up residence in the cloud forest.

It is a weird enough experience to fly above the clouds in an aeroplane, but to walk through a forest and then suddenly emerge above the clouds, all under your own steam, is quite a shock. You can even walk with your head in the hot sun and the rest of you immersed in cool, clinging mist.

The upper cloud forest is marked out by the presence of the only large gymnosperm that is a true native of Ecuador. Its name is *Podocarpus oleifolius*, which does not signify that it is of Spanish descent (olé) but that its leaves are shaped like those of the olive tree (*Olea*). These small, tough, shiny leaves are adapted to minimize water loss for once above the main bank of clouds the climate is much drier both because of lower rainfall and mist and due to the fact that water loss by run off is very rapid from the steep mountain slopes.

It is at this level that the bromeliads really come into their own as cliff hangers, the reservoir of water that is stored in the base of their leaf rosette tiding them over any short periods of drought. This is one reason why the bromeliads make such good house plants and why the growing instructions always say water into the funnel of leaves and not into the soil. It is their own personalized reservoir that keeps them growing in places where no soil will ever form. On the high exposed cliffs it is easy to see the bromeliads displaying one of their weirdest characteristics–they advertise the fact that they are about to flower. When young the leaves of many of the species are green, often with inconspicuous red veins. As they mature, and before the onset of flowering, the leaves become redder and redder, which attracts would-be pollinators in search of nectar. When at last the spike of flowers arises from the bowl of red leaves, the humming birds and insects are there ready to get on with the job. In this way the plant may save on energy rich nectar. In many other plants it is the young leaves that are red and in these cases it is thought that the pigment protects the delicate new chlorophyll from damage by too much light.

The drier conditions of the higher levels of the cloud forest have far-reaching effects on the flora. Every component of the forest appears to be more able to tolerate periods of drought, leaves are smaller, their surfaces covered with a thick shiny cuticle and the stomatal pores are restricted to pits or grooves, where they are often covered by a weft of hairs. Epiphytic flowering plants, ferns, mosses and liverworts are fewer, their place being taken by beard lichens.

There are, however, even at this level (around 3000 metres, 9840 feet) some big botanical surprises, not the least being giant plants, some as much as 5 metres (16·4 feet) tall, that belong to families like the daisies and the poppies which most temperate botanists tend to think of as always being tiny herbs. One giant among the cloud forest leaves are those of the spiny rhubarb plants, *Gunnera*, some of which are more than 2 metres (6·5 feet) across. They grow on the bases of cliffs where water and nutrients tend to collect, and they make ideal water garden plants at any latitude where there is not too much frost. Apart from its massive leaves and spike of flowers, *Gunnera* has a hidden secret which in part accounts for its success, at least in size. In the tissue at the base of the giant leaf stalks are delicate colonies of blue-green algae which are very similar to those primitive plants that were among the first products of evolution. Like many of their forebears they have the ability to fix atmospheric nitrogen turning it into high-class nitrate fertilizer. The *Gunnera* thus provides a sheltered home for the algae which in turn provide their landlord with fertilizer. They live together in symbiosis, a state of mutual help, and together they play their own special part in the cloud forest.

In the absence of accurate measurements all one can do is make a guesstimate concerning the diversity and the standing crop of the cloud forest to allow comparison with that given above for the lowland forest. You would be very lucky to find more than 300 vascular plants in any comparable 2 square kilometres (0·8 square mile) of cloud forest, and as far as standing crop is concerned a figure of 20 dry kilograms (44 pounds) holding 60 litres (13·2 gallons) of water per square metre (10·7 square feet) would not be too conservative an estimate.

Above the tree line the plants really do need any help they can get, be it from symbionts or simply by growing in sheltered locations, for

Overleaf The commercial break. The bromeliads are members of the Pineapple family or Bromeliaceae and here is one indulging in a lot of advertising. The red leaves attract pollinators long before the red flowering spike comes up in the middle of the bowl of leaves.

Above Ecuador has a lot of different trees and a lot of different tree frogs, many of which still wait to be discovered. The rain and the cloud forests make an ideal habitat for both the 'amphibious' ferns and the true amphibians. There is plenty of water to keep their skins moist enough to act as an all-enveloping lung and lots of pools in which to croak their love songs and lay their eggs. The hylid tree frogs of South America are closely related to those in Australia. Since frogs can not live in salt water and could not hop that far this is good evidence of continental drift.

here on up the conditions for growth get much tougher. Just because the tree line is easy to recognize does not mean that it is easy to explain and the theories as to exactly what factor controls the growth of trees are almost as numerous as the workers who have proposed them.

At first the trees simply move down into the shelter of the valleys where they receive both shelter from the wind and any excess rainwater draining downslope, the intervening ridges being covered only by dense scrub about 2 metres (6·5 feet) tall. Higher still the pockets of forest disappear and the scrub gets more and more stunted and much sparser in distribution. This is called locally the wet paramo, a name which has been adopted by the scientific community to describe this type of upland vegetation.

Above the tree line conditions are certainly much drier than in the cloud forest. There may be a frost on any night of the year and 30 centimetres (12 inches) down the soil records the fact that the mean annual temperature is below 9 degrees Centigrade (48·2 degrees Fahrenheit). However, the surface layers of the soil may warm up each day to temperatures that are considerably higher, especially in areas where the soil is dark in colour, for all black bodies absorb heat very efficiently. Although, as we shall see, the plant shoots can protect their delicate growing parts safe inside buds or hidden among leaves, roots are not so lucky. One function of the root is to keep the plant supplied with water and minerals and to accomplish this delicate living cells called root hairs must be exposed to the soil environment. In the same way the root tip must be able to grow down, pushing its way between the soil particles, and a protective covering even of scale leaves would be an enormous hindrance. It is for this reason that, although the roots of plants live in a much more stable environment than the shoots, they are still very sensitive to temperature. If the sub-soil is too cold then all a plant can do is turn its root tips up and exploit the zones near the surface where, at least during the day, it is warmer. A shallow root system means that the plant must exploit a larger area of soil from which to draw its water and minerals; hence, at least in part, the scaling down of the vegetation of the paramo both in terms of density and height.

It is across the lower zones of the paramo that the tropical lowland element of the flora finally disappears. Perhaps one of its last stands is seen in the diminutive bromeliad *Tillandsia usneoides*, the Spanish Moss, that festoons every available bit of hanging space. This is a plant that can really be said to live in thin air, drawing its minerals in part from the dust that is washed on to it by the rain. In some of the villages on the paramo this weird plant festoons the telegraph wires, becoming extra luxuriant at the cross roads where more dust is stirred up into the air.

In comparison with lowland forests, a diligent search in a 2-square kilometre (0·8-square mile) area of bush paramo would be fortunate to

Spanish Moss festooning a tree. This diminutive epiphyte is another member of the Pineapple family. It is called *Tillandsia usneoides* because it looks not unlike *Usnea*, an epiphytic lichen. It is of interest that in the drier parts of the cloud forest the very hardy *Usnea* lichen, commonly known as old man's beard, takes over as the dominant epiphyte. Though tolerant of drier conditions, *Usnea* disappears at the hint of atmospheric pollution.

reveal 60 species, with 5 dry kilograms (11 pounds) holding perhaps 1 litre (0·22 gallon) per square metre (10·7 square feet).

It is of great interest to speculate on the fact that all this has happened in the last 90 million years. Before that time there were no Andes and the mighty river Amazon flowed eastwards into the Pacific. At that time the whole of this part of South America was lowland tropical and was covered by what is best termed a neotropical flora. As the Andes have slowly risen up turning the Amazon in its tracks to flow into another ocean, parts of the tropical flora have gradually been exposed to the cooler and cooler climates of higher altitudes and evolution has been forced to happen. Those plants whose genetic make-up was not sufficiently varied or plastic to allow adaptations were left at the starting gate down in the lowlands. The others rose up with the mountains, their communities being bolstered by immigrants from cooler climes in higher and lower latitudes.

The best way to understand the necessary adaptations is to take a close look at any piece of unspoilt paramo. The shrubs are all characterized by small, tough leaves, often with a covering of dense white hair. Their stature reflects the amount of protection that their habitat provides from the wind. In the least protected places and at the highest altitudes, the shrubs are replaced by much lower growing plants. These include shrubs which form mats of branches close to the ground in which all the leaf litter collects producing a layer which protects the young buds from the nightly frosts; cushion plants that are so tightly adpressed to the soil that hardly a leaf shows above their contoured cushions; and tussock grasses with tightly rolled leaves thus greatly reducing the leaf surface area from which water can be lost. Most striking of all are the rosette plants and especially the rosette trees, which in undisturbed sites may be as tall as 10 metres (33 feet). Their trunks are covered with a thick cloak of old dry leaves or leaf bases and the gigantic sword-shaped leaves that form the terminal rosette are often covered with a dense felt of hairs.

It is very easy to jump to the wrong conclusions concerning adaptations and many of the features described above would be of equal use against grazing; it must be remembered that the high Andes are the native home of the Llama, Alpaca and Vicuña, all members of the camel family and all voracious nibblers.

One of the greatest mysteries of these paramo plants is what makes them decide to flower. There is no real variation in day length, which is often an important factor inducing flowering in temperate plants. In those areas where there is a wet and a dry season the stimulus may come from the change in pace of growth related to water supply. In such situations, the main period of flowering often coincides with the driest period, when nightly frosts alternate with the hottest days of the year.

Most surprising of all is the flowering of the larger rosette trees. It would appear that they sit tight for many years and, in the case of the large *Espeletias*, or tall friars as they are locally known, perhaps for more than 100 years. When they are ready they put all their mature energy into the production of a massive spike of flowers and after the seed is set they die.

The Pineapple family which has followed us all the way up from the lowlands in the shape of the bromeliads and *Tillandsia*, is not to be outdone and is present in the paramo as a rosette tree, the puya. Each short trunk is terminated by a rosette of fleshy leaves, each one armed with vicious down-curved prickles. The leaves not only protect the central bud from all would-be grazers, but form a deep cup which warms up very quickly once the sun shines into it. As the flower spike begins to grow and the flowers start to open, the hummingbirds come in search of nectar and with them comes the puya mite that hitches a ride in the nostrils of the tiny bird. As the flower spike extends, the older flowers fade and new ones open, and the mites migrate from one to another until they are all concentrated in the last few flowers. They then seem to know that it is time to leave. The next hummer in is the last flight to a new food supply and they all leap on board.

The hummingbirds of the high paramo are themselves faced with a real problem. They use up a lot of energy buzzing around to feed during the day and each night they must use up a lot more simply to keep their diminutive bodies up to the correct temperature. In order to reduce heat loss the hillstar hummer roosts in small caves. It is, however, impossible for them to overcome their territorial behaviour pattern and so they do not huddle together to keep warm but roost neatly spaced out on the walls where they are at least protected from the frosty winds.

Proceeding higher, not even the hillstar can make a living and the plants get smaller and smaller until only one flowering plant is left–*Culcitum nivalis* which, like the tall friars, is a member of the daisy family which appears to have produced some really tough customers the world over. *Culcitum* grows flattened to the ground, its long leaves covered with a thick down of white hairs. However, a close look shows that it does not rule its penthouse roost alone for growing among the stones are a number of lichens, mosses and algae, although the mass of the average total standing crop would not be worth measuring.

At around 4800 metres (15 740 feet) permanent snow and ice cover the volcanic peaks in a blanket that is always at or below the freezing point of water–an ice desert on the equator, but even here there are bacteria that eke out an existence living in the melt water between the crystals of ice. The three mighty volcanoes, Chimborazo, Cotopaxi and Cayambe measured from the centre of the earth are the highest mountains in the world. These three at present lie dormant whereas Sangay to the south which is considered to be one of the most active volcanoes in the world exuberantly lives up to its reputation. Whether sleeping or active the volcanic peaks of the Andes all tell the same story. This is a young range of mountains still in the making, pushed up by the meeting of two pieces of the earth's crust which are each going their own ways across the face of the globe.

The young peaks contain the eyries of the Giant Condor (*Vultur gryphus*) and from their flight paths not only must they get a panoramic view of

Far left A puya in full flower standing among a crowd of admiring tall friars. The former is a member of the Pineapple family, the latter a member of the Daisy family. Both have the growth habit of the rosette tree, which includes a short trunk, a terminal rosette of leaves and a once-and-for-all gigantic spike of flowers. This is a case of convergent evolution – two unrelated plants fitted to do the same job.

Left This hummingbird caught in mid-hum is one of the many hundreds of different birds that live in the South American rain forests. They are among the luckiest birds in the world. The stable climate provides food for 365 days of the year and so they have no need to migrate. Their only real problem is in manoeuvring among branches of the canopy where most of their food is to be found. However, the hummingbird, which looks more like a giant insect than a bird, has short wings which are designed for accurate flight while refuelling, so getting round branches provides no difficulty at all.

Left Culcitum nivalis, the first or the last flowering plant on Mount Cotopaxi, depending, of course, on which way you decide to approach it. I can assure you that at that altitude it is much easier going down. *Culcitum* is another member of the Daisy family, Compositae, which enjoys success on a worldwide scale. Comprising more than 14 000 species, moulded into almost every ecological type and life form, they beautify our gardens and provide us with some of the most noxious of weeds.

Overleaf Mount Cotopaxi, 5850 metres (19 500 feet) high, is just one of a long line of active volcanoes all of which wear a deceptive cap of snow, at least when not actually in eruption. They are part of the longest mountain range in the world (above water, that is, see p38) and form the backbone of the continent of South America. When in eruption they not only melt the ice cap with catastrophic results for those in the way of the flood, but also show that both continental drift and mountain building are still happening.

the volcanoes but also of all the zones of vegetation. In no case will they see a single super plant but a super community of plants on which they ultimately depend for their livelihood.

From the air it is possible to see how many factors other than altitude affect the basic zonation. Most significant of all are the rain shadows produced by various peaks which bring periods of drought to certain regions. The effect of water stress under the tropical sun is much the same at each level and ranges from a short season when certain plants lose their leaves through semi-desert conditions, with succulents and cacti dominating the vegetation, to true desert in which no large plants can grow. The severity of the effect is related to the duration of the dry period.

The same is true the world over, the vegetation of any region reflecting the potential of that particular environment for life. If you know your plants and understand the vegetation you can read the environment like a book.

But why complex communities, why not single super plants? Imagine that you could select 2 square kilometres (0·8 square mile) near the Rio Palenque Reserve and strip it of all its plant and animal life. Imagine too that all the different sorts of plants that evolution has ever produced are there on the outside ready to colonize your newly bared plot. (It is not such a strange idea, in fact every time you clear a plot in your garden you are performing a similar experiment.)

Now begin to act as a highly conservative botanical Noah and let the plants into your ark one by one. Certain of the plants would in all probability be unable to grow because it was either too hot, too wet or perhaps the bed rock and hence the soil would be of the wrong type. The first one that could grow would come in with true pioneering spirit and would fill the space using some of the minerals, some of the water, some of the light, some of the potential. A pure monoculture, it sounds good, but problems would come as soon as all the nitrate was used up because there would be no blue-green algae to fix any more, or when the leaves fell off at the end of their life span there would be nothing to break them down releasing the minerals for future growth. So you would have to let more plants and animals come into the plot until all the niches were filled and a new living system became stabilized. Then and only then would the full potential of the plot be realized.

Each living system the world over consists of the following components – primary producers, consumers, decomposers and parasites. The primary producers are, in 99·9 per cent of cases, plants that contain chlorophyll and, therefore, have the ability to trap and fix the energy of the sun. Consumers are herbivorous animals that eat plants, carnivores that feed on other animals and omnivores that eat just about anything they can get hold of. Decomposers live on dead and decaying organic matter, releasing and recycling the minerals for re-use by the system. Parasites feed on living animals and plants causing diseases in their hosts. Each separate organism has been fitted by evolution to play its own particular role but it is always a role in a complex system.

The conservative Noah experiment would be impossible to perform simply on account of the time that would be required. Nevertheless, it is interesting to speculate on the outcome. Would you succeed in producing a new type of living system, a new type of vegetation, or would you end up with something that was in essence the same as that which is found in the Rio Palenque reserve? The answer is in all probability the latter for it is unlikely that you would be able to improve on the end product of many millions of years of community evolution.

It was in the late 1950s that biologists the world over got together and decided upon a programme of research, part of the aim of which was to study the status quo of the vegetation of the world, at least in terms of the amount of energy it fixed, stored and used each year.

It was a massive undertaking for not only was there a lot of ground to be covered but many of the techniques had to be developed, perfected and standardized for international use. Suffice it to say that eventually the job was done, and the International Biological Programme pro-

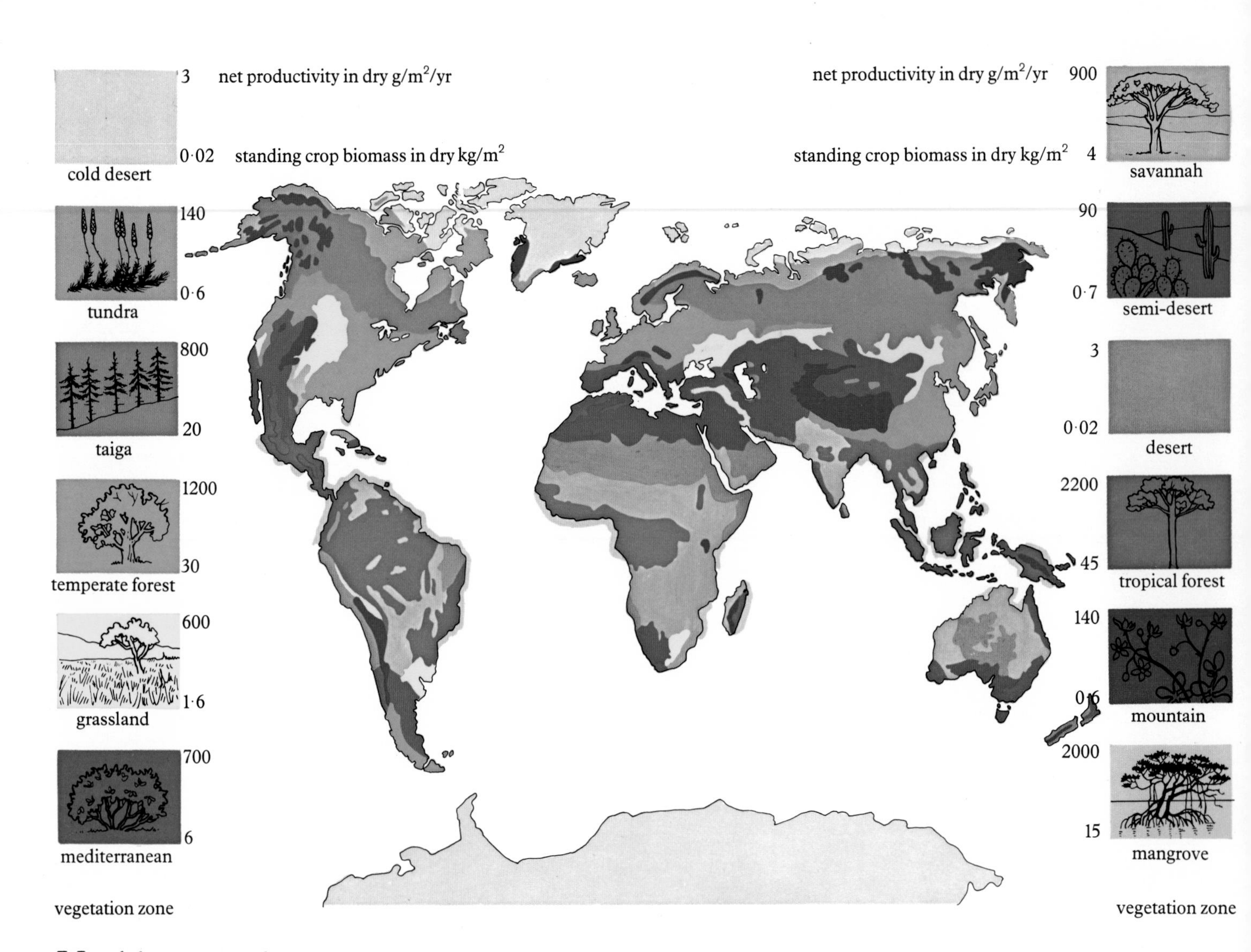

World vegetation

The early naturalists who travelled the earth on foot and on horseback realized that the vegetation grew in distinct zones which in broad terms reflected the climate in which they grew. This map is simplified from that produced in 1969 by two ecologists, a Russian, N. I. Bazilevich, and an American, J. Olson, who pooled their knowledge and some of the results from an international research effort aimed at understanding the relationships between climate, vegetation and productivity.

duced the first meaningful figures for the amount of energy fixed in unit time by a cross-section of the natural vegetation of the world.

The rough figures are shown on the map (above), which is a gross over-simplification because it does not show any variation due to bed-rock type and, hence, availability of nutrients. Nevertheless, it is possible to say that the productivity of land vegetation is in the main related to the mean annual temperature and the distribution and the amount of rainfall.

The total amount of energy fixed by all the primary producers of the world is $3{\cdot}5 \times 10^{20}$ joules ($2{\cdot}6 \times 10^{20}$ foot-pounds) per year and incredible as that may seem, it represents an efficiency of only 0·01 per cent. It has taken the process of evolution more than 3400 million years to become that efficient.

The fascinating thing is that wherever you look, from the anchovies of the Humboldt current, living on their plankton biscuit 0·2-millimetre (0·008 inch) thick, through the 80-metre (262-foot) tall tropical rain forest of the Rio Palenque to the high paramo of Mount Cayambe you find a complex system of plants and animals each playing their own roles to the perfection allowed by their own state of evolution, all members of an integrated productive community. In addition, each one of them is made up of a 'mixture' of some of the same twenty-two elements borrowed temporarily from the environment.

Evolution has moulded the living chemicals into about 1·5 million different forms of life, fitting them to the opportunities on offer. Community evolution has gone hand in hand with the process, providing more and more opportunities, maximizing the flow of energy.

Matter/energy can neither be created nor destroyed, it can only be changed from one form to another and at each change there is some degradation, all systems tending to degrade the energy associated with them. These are the basic laws around which the universe revolves. Evolution had to happen and responding to these laws it has taken over a dead planet, covering it with a film of life best called the biosphere and, inefficient as it may be, it is all that we have got.

One of the reasons behind the International Biological Programme was to ascertain the potential of the planet earth to man and there is certainly no better place in which to do just that than along the transect from the coast to the top of the mountains of South America. A visit to one of the many colourful markets of the Andean slopes is not only a fascinating experience but also a lesson in economic botany. The stalls, if such you can call them because they are all situated at ground level, are overflowing with all the products of the various zones of vegetation–banana, pineapple, papaya and cassava from the lowlands; citrus fruits from the drier regions; carrot, cabbage, tomato, potato (there are 421 different sorts), strawberries, peanuts, peppers, lima beans, maize, wheat and many more all grown at the higher levels above the natural tree line. Add to this a range of spices, flavours and dyes to tempt the menus and the fashion houses of the world and that is real productivity for you. Much of the local commerce is still carried out by barter in which you swap your excess for what you need and vice versa, each farmer playing his own role within the total system.

In some ways the farmers of the tropical Andes are among the luckiest in the world. For 12 hours each day the sun shines on their fields and for 12 hours darkness reigns, so if they do not try to grow their crops too far up the mountain every day is a potential growing day. By adjusting their life styles to fit in with the terrain and the climates of the zones, they are able to grow just about anything the world has to offer. Even the highest snow-free slopes can be used for grazing Llama, Alpaca and Vicuña, which not only provide meat but also the softest warmest ponchos in the world.

The fertile valleys like that of the Urubamba, a headwater of the Amazon and once the sacred valley of the Incas, have been keeping a considerable population well fed since man arrived in South America.

It is well to remember that before South America was opened up by the new explorers the world markets had much less to offer than they do today. The slopes of the Andes and the lowland forests on either side have provided the world with such important natural products as potato, tomato, maize, ground nut, lima beans, cassava, papaya, passion fruit, peppers, vanilla and so on. At one time these were all members of the local natural flora, and today they are grown in other parts of the world which enjoy what are in essence similar climates. Take all those out of your weekly shopping list and you have not got much variety left.

The contemporary importance of these on a world scale can not be overstressed. Apart from the staples and the flavours that have been injected into the diets of the world, this part of South America has provided tobacco; coca for the world's most sought-after cola; guano, the fertilizer that helped to feed the industrial revolution; rubber to tyre the motor industry; quinine, the first known antimalarial that opened up much of the humid tropics to modern man's endeavours; and nasturtiums, fuchsias, and calceolarias to grace the gardens of the world.

I wonder what else it has to offer. And I wonder whether we will ever know because at the moment the rate of destruction of the natural vegetation is much faster than the rate of its study. But I am jumping the gun. Long before man first stood upright somewhere in the great rift valley of East Africa evolution had completed the job, it had conquered the world. Apart from the ascent of man, the last 2 million years have seen great changes in the natural environments of this planet, and those changes have had far reaching effects on the vegetation and on the course of evolution.

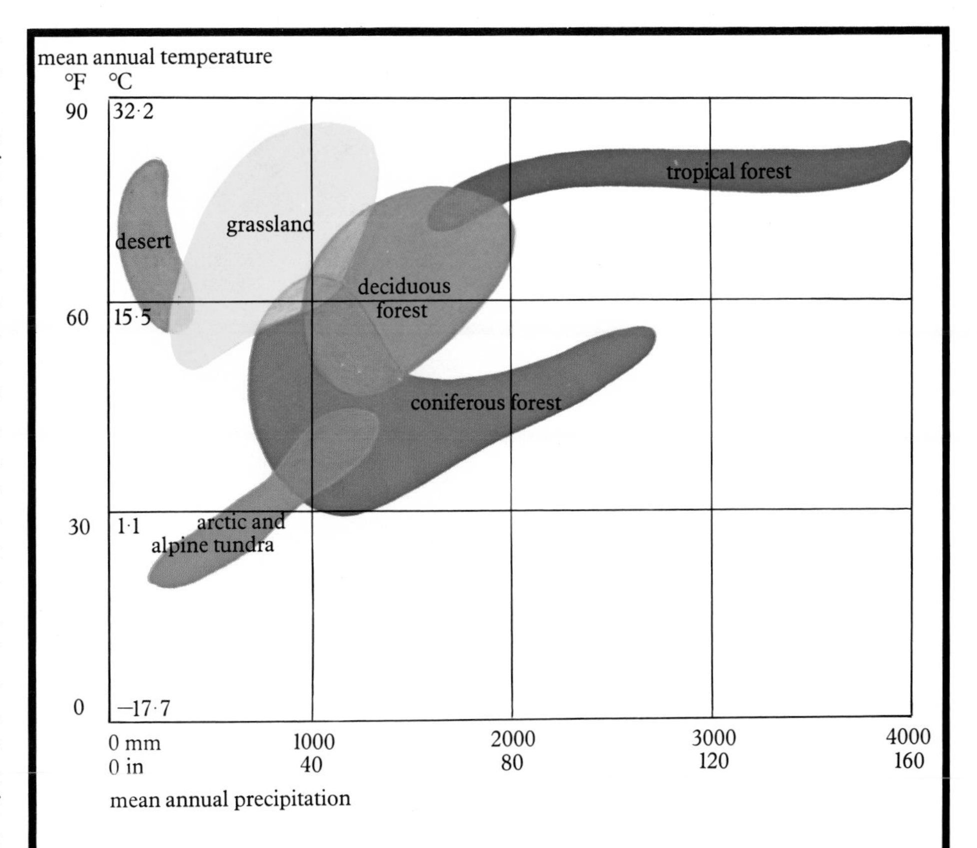

Vegetation, climate and productivity

Due mainly to the work of Heinrich Walter (see p6) it is now possible to understand the major climatic factors which determine the distribution and the potential productivity of the six major types of world vegetation. All you need to know is the mean annual temperature and rainfall of your own back yard and you can then find out from the diagram the type of stable vegetation which was there before your back yard came into existence. It will not work if you happen to live in an area in which seasonal distribution of rainfall is the main factor which determines the type of vegetation.

Overleaf An Andean market overflows with the produce of many local climates. The skilled Andean farmer can grow a great variety of crops by using the diurnal climate to the best advantage. The main problems are prolonged periods of drought which may be overcome by irrigation, and the cold winds of frosty mornings which are overcome by a poncho of finest Alpaca wool.

Chapter 7

White death, new life

On at least four occasions over the last 2 million years, the ice sheets of the polar regions have expanded, obliterating all life in their paths. As the climate improved the white death receded to be replaced by a new mantle of living green. A journey from the north pole down through Canada shows what the stages in this recolonization may have been.

During this time another important event was occurring, the effects of which are more widespread than any of the ice ages – the ascent of Botanic Man was underway.

Water is the commonest substance on the face of the earth and it can exist in four states–as a solid (ice), a liquid (water), a gas (water vapour) and as life (the products of evolution). At present the water resources of our planet are distributed as follows–almost 98 per cent liquid, 2 per cent solid, 0·001 per cent gas and 0·0001 per cent life. We have seen in the previous chapters that the amount of living water has been gradually increasing throughout evolutionary time and that it is at present distributed across the globe in amounts that reflect the potential of each environment. On the land, the major factor limiting life relates to the cycle of change of water from the liquid to the gaseous state–the cycle of evaporation and precipitation.

There is, however, another change of state of the earth's water resource which has, over the past 2 million years, had a massive influence on the course of evolution, and that is the change from liquid to solid. We know that on four occasions over that comparatively short period of time the ice sheets of the polar regions have expanded, obliterating all life in their paths. Only the warmer latitudes, and especially the tropics, escaped the ravages of the great white death and even there, the permanent snow caps of the high peaks pushed down their own icy fingers of destruction.

Exactly why the various glacial and interglacial periods happened is still a matter of heated debate by palaeoclimatologists (people who study past climates). The fact that during this period the continents have been slowly easing themselves into their contemporary positions, and the current phases of mountain building and erosion have been shaping not only the modern atlases but also the modern climates is probably enough in itself to account for the vagaries of glaciation. It is still fun to speculate on the possible effects of sun-spot cycles, greenhouse effects or even of integalactic warfare. The indisputable effects of the ice and its movements are, however, there in the glaciated landforms, ice-worked sediments and in the interglacial and postglacial deposits of peat. The latter, though not on such a grand scale as the coal of the Carboniferous, are, in their own way, packed full of information concerning the history of the most recent period of geological time, the Quaternary.

As the glaciers built up, so the level of the world's oceans went down as the cycle of evaporation and precipitation was frozen. At maximum glaciation an extra 6 per cent of the world's water became solid and the general level of the oceans fell by approximately 100 metres (330 feet). The marine animals and plants were fortunate because the vast majority of the inshore forms could migrate with the slowly receding tide to take up similar habitats in the new littoral and sublittoral zones. In contrast, on the land whole zones of vegetation were wiped out and migration of the mobile forms was difficult because the glacial hinterlands were already occupied by their own characteristic living systems. It must be borne in mind that the developing ice sheets did not push back the climatic zones in a regular fashion. If a glacier ended in what used to be a piece of cloud forest, the conditions for the continued growth of that type of vegetation would have continued to exist not all that far from the snout of the glacier.

As well as the major effects of destruction the ice sheets also performed an important task of reconstruction. Whole landscapes were reshaped, mountains were cut down to size, new valleys were gouged out, old ones were scoured into a U-shaped form, rivers and streams were dammed to form lakes, and glacial debris was carried down towards the sea where it was deposited in the form of great deltas. Perhaps the most important effect was that bed rock of all types was ground down and in the process supplies of minerals that had been locked up for millions of years were released in an available form.

During the glaciations the living world lost much ground, at the most about 20 per cent of the total land surface area, but all that new potential was there 'waiting' for the postglacial spring that would release the water and the land from the bondage of ice, making life possible once more. So it was that as the climate began to improve, the white desert was slowly replaced by a new mantle of living green. Today it is possible to take a trip from the North Pole down through the Canadian Arctic to the warm southlands of latitude 39 degrees north and see all the possible stages of recolonization held in 'cold storage' by the zoned environments of the great continent.

Passing north or south from the tropics is somewhat akin, both in climate and vegetation, to climbing up from the coast of Ecuador to the top of Mount Cayambe. In essence, the further you travel the lower will

Overleaf Arctic spring at Galbraith Lake, Brooks Range, Alaska. The ice caps of the high mountains stay throughout the year, acting like a giant refrigerator, cooling the winds which blow across their flanks. Lower in the valleys spring brings the melting of the surface ice and the promise of warmth and enough liquid water to make the tough but beautiful plants begin to bloom.

Below A U-shaped valley produced by glacial action. I well remember being shown my first such valley in the peace of the English Lake District and wondering just how ice could have scoured away such an enormous amount of rock. It was not until I stood on a real live glacier, almost surrounded by deep crevasses and the noise of slowly moving ice (or did I imagine it?), that I understood the power of solid water.

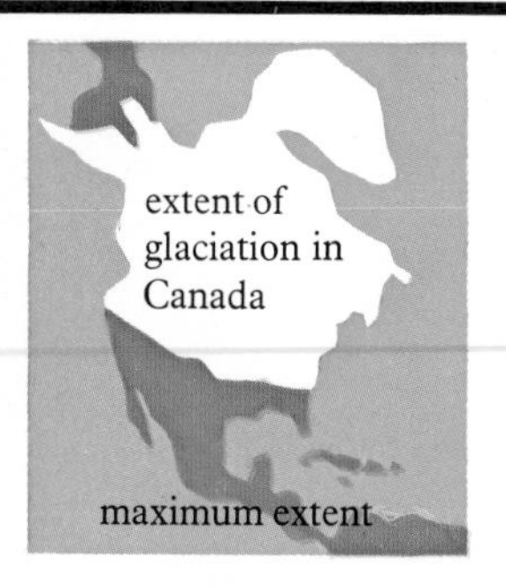

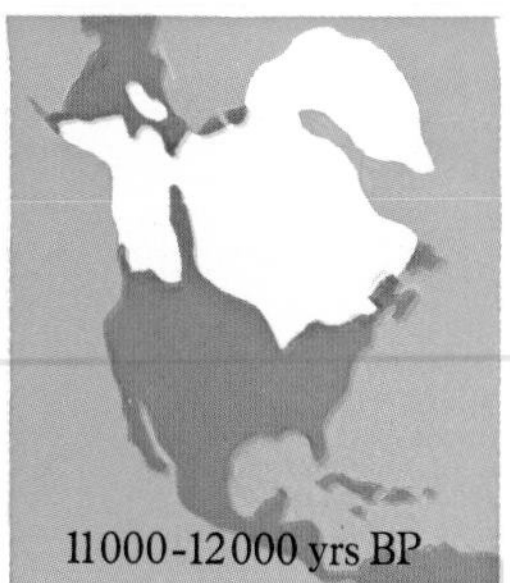

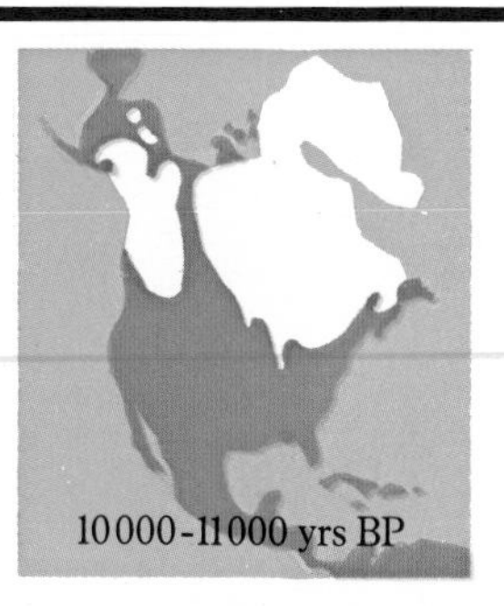

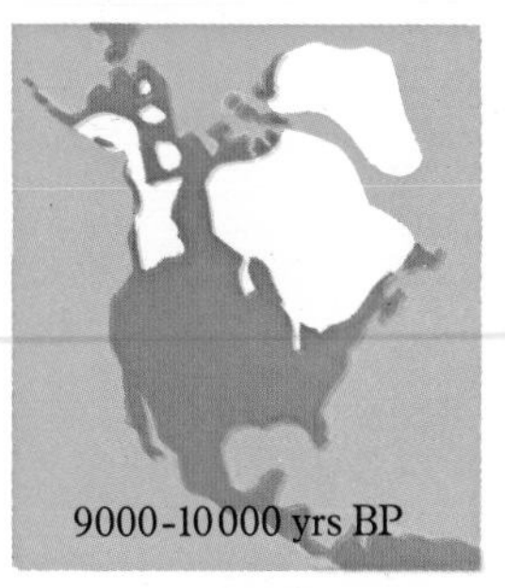

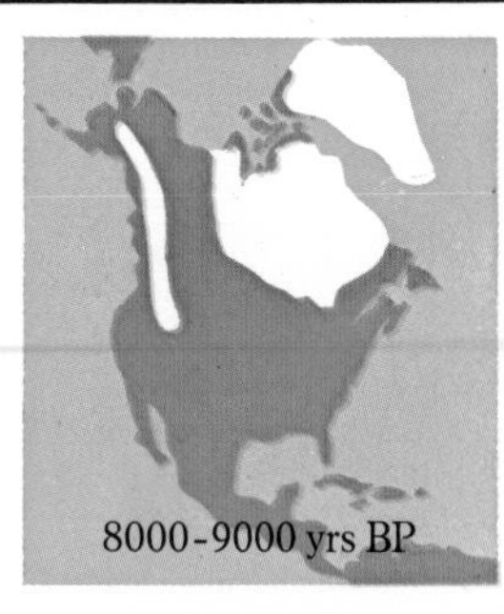

Canada

Above The end of the Wisconsin glaciation opened up the New World for recolonization by plants and by man. It is of interest that much of Alaska was free from ice during at least the latter part of the glaciation and an ice-free western corridor rapidly developed during the late glacial period. Perhaps most important of all was that a land bridge was in existence across the Bering Straits up until at least 10 000 BP giving a direct land route for the immigration of man.

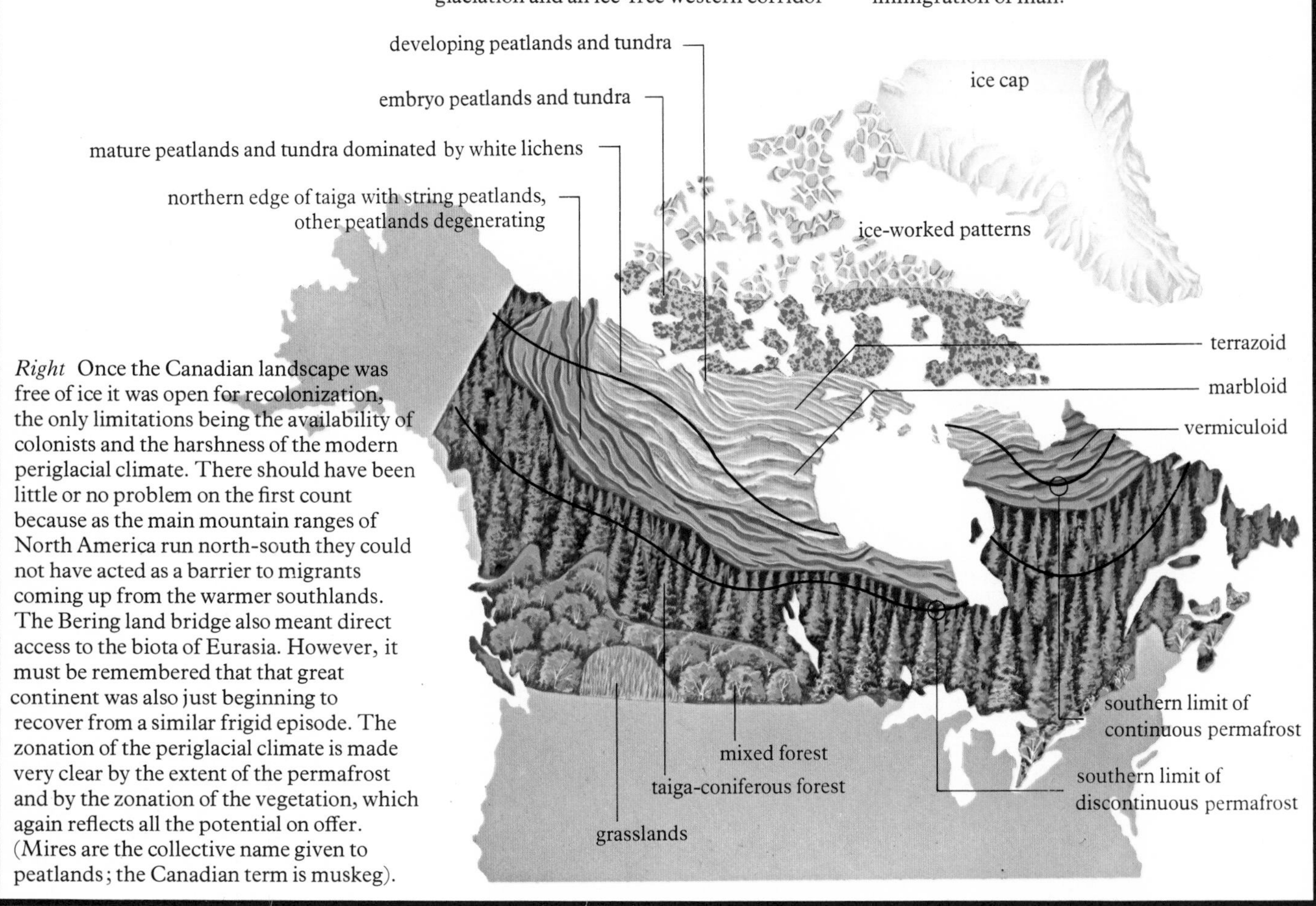

Right Once the Canadian landscape was free of ice it was open for recolonization, the only limitations being the availability of colonists and the harshness of the modern periglacial climate. There should have been little or no problem on the first count because as the main mountain ranges of North America run north-south they could not have acted as a barrier to migrants coming up from the warmer southlands. The Bering land bridge also meant direct access to the biota of Eurasia. However, it must be remembered that that great continent was also just beginning to recover from a similar frigid episode. The zonation of the periglacial climate is made very clear by the extent of the permafrost and by the zonation of the vegetation, which again reflects all the potential on offer. (Mires are the collective name given to peatlands; the Canadian term is muskeg).

the mean annual temperature of the area be. Unfortunately, once outside the tropics it is not possible to rely on the subsoil temperature as approximating to the mean because the further you are from the equator, the more is the diurnal climate replaced by a seasonal climate. This means that any spot check using a soil temperature probe will simply record the march of the seasons. In the higher and lower latitudes there is only one way to obtain a mean figure for the annual temperature and that is to set up a recording meteorological station and at the end of the year do your homework. As far as the plants and animals are concerned this would be a useless exercise, because in a seasonal climate annual means are of little importance compared to the length of the growing season– that is, the period of the year when the green plants can grow. The exact temperature at which plant growth starts and stops varies from species to species. However, as a rule of green thumb, the growing season is often taken to be that part of the year when the mean daily temperature is at or above 7 degrees Centigrade (44·6 degrees Fahrenheit).

The warmest place in the whole of central Canada is on the Niagara peninsular where 170 growing days are enough to ripen the grapes that form the basis of all the best Canadian vintages. One of the reasons for Niagara's long growing season is the presence not of the largest falls in the world but of the world's largest lakes, the waters of which lose their summer heat much more slowly than the surrounding land. The Great Lakes thus act as a giant storage heater which greatly ameliorates the temperature of the early part of the autumn.

Camera crew on the edge of the arctic ice cap. Bill Radforth is in the foreground.

Passing north, away from the similar warming effects of the maritime areas, the number of growing days decreases rapidly, as does the length of time since the land was under a burden of ice. At its maximum, the North American ice cap covered much of Canada and stretched well down into what are now the United States. Since this time the ice has been melting, revealing more and more new landscapes ready for colonization.

In the far north the idea of a burden of ice becomes very real for there the earth's crust is still recovering from the immense weight of the ice and is actively rising to the occasion. In places this rate of recovery is so fast that it is possible to find the skeletons of marine creatures, like whales, which must have been stranded on a beach now many tens of metres above sea level.

Apart from its elastic crust, the vast northlands hold many secrets, none of which are more extensive than the permafrost. The glaciers may have gone, but beneath the rolling monotony of the high tundra there is a continuous sheet of ground ice, only the surface layers of which melt during the growing season. Throughout the winter the ground is frozen solid right up to the surface, but as spring advances the surface layers begin to melt, producing what is known as the active layer. The active layer is charged with water which bubbles up from the melting ice and it is in this that any developing vegetation will have to find root space, because below this the water is solid and is thus not available for the growth of plants.

The term 'active layer' is not derived from the activity of the roots, but from the activity that is caused by the annual cycle of freeze and thaw which really stirs things up in a big way. The glaciers leave behind rock debris, which is made up of bits and pieces of all shapes and sizes. Wherever the terrain is more or less flat, the spaces in between the ground rock are filled with water and that is where all the action comes from.

Anyone who has had the misfortune to experience a burst pipe will know to his or her cost that when a certain volume of water turns to ice it expands. It is this expansion that fractures the pipe, but the crack is only revealed once the ice has melted and the water comes pouring out. Annoying as a burst pipe may be, this phenomenon of expansion and contraction is one of the most important properties of water, for without it there would be little or no life anywhere on earth.

If you fill a kettle to the brim with water and put it on to boil, the volume of water which once filled the kettle will soon not only fill your kitchen, but also much of your house which will get all steamed up. You will still have the same number of water molecules that as a liquid filled the kettle, but now as a gas they occupy a much larger volume. This form of expansion is the basis of the steam engine in which part of the energy that pushes the molecules apart is converted to push the train along. In a modern triple expansion engine the efficiency of energy conversion may be as high as 5 per cent. In the steam engine the steam is allowed to condense back into water but unfortunately you cannot get the energy back, because it is now spent energy having been converted into another form. (Remember our basic rules, page 28.)

Having got all your steam back in the kettle now put it in the freezer and allow it to cool down. The water molecules begin to pack down closer and closer together and for this reason the volume of water in the kettle will decrease ever so little down to the 'magic' temperature of 4 degrees Centigrade (39·2 degrees Fahrenheit) when lo and behold it begins to expand once more. Most solids, including ice, have a rigid crystalline structure. Ice has a very open structure, so much so that ice is actually less dense than water. This is unusual, for most substances are denser as a solid than as a liquid. Below 4 degrees Centigrade ice crystals start to form and the water becomes less dense. The ice floats to the top and the denser 4 degrees Centigrade liquid falls to the bottom. In the case of your kettle, the process will continue until all the water is frozen and, because of the increase in volume, a column of ice will soon protrude from the spout.

Water

The three faces of water. If you want to see the fourth take a look in a mirror, the garden pond on a still day will do.
Above left In steam, the water molecules move about independently, travelling at enormous speeds in random directions. The hydrogen atoms have a small positive charge and the oxygen atoms have a small negative charge. *Above right* This produces an electrical attraction between neighbouring molecules which in liquid water causes them to cluster together. This is known as hydrogen bonding. In liquid water the hydrogen bonds are continually forming and breaking, producing an ever-changing structure. *Below* In ice, the molecules are arranged in a three-dimensional tetrahedral structure. When ice melts, this open structure collapses and the density of water increases.

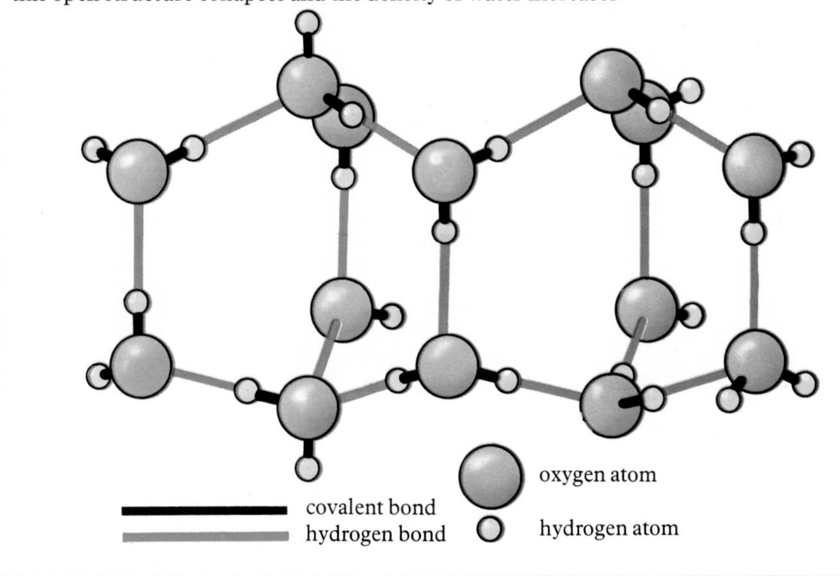

It seems simple, but just think what would happen if ice were denser than water. Instead of floating it would sink to the bottom and the sea would effectively freeze from the bottom upwards. A solid sea would have certainly put paid to much of evolution at the beginning of Chapter 3.

Due in part to the insulating properties of the soil, the ground ice permafrost behaves in exactly the opposite way to a body of open water. The ice remains at the bottom and only the top thaws. All the effects of expansion and contraction thus take place in the surface layers, the effect being to sort the larger stones from the finer gravels and silts, producing remarkably symmetrical patterns.

On gentle gradients, the glacial debris is aligned into series of ridges and furrows that run predominantly downwards. These stone stripes braid out on flatter terrain, where they lose themselves in a matrix of stone circles and ice wedge polygons. The former abound in areas where stones and pebbles predominate in the substrate, while the latter typify terrain that is composed of finer sediments. Stone circles are usually quite small while ice wedge polygons range from a few centimetres to more than 100 metres (330 feet) across. The largest ice-worked patterns of all are called pingos and are often found in estuarine situations, being built up of fine silts. As a pingo grows upwards, the top begins to collapse, eventually forming a volcano-like structure, although pingos, unlike volcanoes, have hearts of solid ice.

The best way to understand the mechanism of all this symmetry is to stand on the active layer and gently rock backwards and forwards on your feet. Very soon the substrata around you turn into a thixotrophic mass and waves move out across the terrain, tilting the stones in their path and rearranging them into new patterns. The trouble is that long before they have attained a new symmetry, you begin to sink down into the semi-liquid mud.

Conditions on the surface of the active layer are extremely harsh for a growing plant. The growing season may only last for a few weeks each year and for almost 6 months the sun is going to disappear completely below the horizon. Fortunately, much of the growing season is blessed by 24 hours of sunshine each day and the air temperature can be pleasantly warm for this short time. The plants which take up the challenge of the highest arctic life, growing within a few metres of the melting ice sheet, must be considered the hardiest pioneers of all. Like all true pioneers they know their terrain and always grow in the

Top Ice polygons with raised centres. A single ice polygon does not look very impressive, but when you cross kilometre after kilometre of tundra all patterned in the same way the whole concept of the active layer of the arctic soils takes on a new dimension. A walk across the polygons in mid-summer not only reveals the instability of the active layer but also the fact that the bulk of plant life starts in the shelter of the depressions around the margins of the polygons. Life on the raised centres is just too harsh.

Above Two large pingos. Pingos come in many sizes but only one shape. At first they appear just like a mound but as they grow larger they take on the aspect of a giant carbuncle. A trip inside a big one is a never-to-be-forgotten experience – a cave carved into solid ice and silt, the walls and roof of which throw back the light from your torch like an all-enveloping chandelier. Such ice caves are man made, never (as far as I know) natural. They are dug for storage purposes and in at least one case to be used as an indoor curling rink.

Overleaf Part of the permafrost exposed at the edge of a stream. Walking across the high tundra the permafrost is never far away, at the most a couple of metres, but it is impossible to see. Here a stream has eroded into a large palsa plateau revealing the lens of ice and the active layer made of purest peat above. Apart from the climate the plants could not have better conditions – throughout the summer melt waters well up from below bringing with them materials dissolved from the glacial debris, a self-regenerating iceoponic garden.

most sheltered spots, which happen to be in the depressions between all those patterns. If you are looking for plants in the periglacial fringe, always land your helicopter on patterned ground. All that ice work in the active layer prepares the ground for the plants to come.

The first footers (I can not call them rooters because none of them have roots) are often blue-green algae and lichens, both of which share the ability to fix atmospheric nitrogen and turn it into useful nitrates. Lichens are dual plants in that they are made up of an alga (often blue-green) and a fungus. The alga fixes the energy both for itself and the fungus while the latter provides a nice tough home for the alga. Alone, neither could live on the surface of the tundra; together, in a bond of mutual help called symbiosis, they bring life to the harshest environments on earth.

With a ready supply of nitrate fertilizer and a little bit of humus, the habitat is complete and other plants can move in. All the newcomers must blow in from the warmer south in the form of spores and seeds, for nothing could have survived the long glacial period under the enormously thick ice.

Among the earliest arrivals are a whole range of mosses, each of which adds its own splash of colour, green, red, orange and gold, to the tundra, in time filling the depressions between the stripes, polygons and circles. It is into the pad of humus produced by the mosses that the first flowering plants find root space. The two which often make the earliest appearance on the Canadian scene are a sedge (*Carex aquatilis*) and the beautiful drooping saxifrage (*Saxifraga cernua*). Both are perennial plants, the former overwintering as an underground rhizome, the latter as bulbils situated at the base of a rosette of leaves. This means that the plants have an extra and very important job to do each summer; not only must they grow, flower and set seed, but they must also store sufficient energy to tide them over the 6 months darkness to come. It must, however, be remembered that like all chemical processes the metabolism of living organisms is affected by temperature. The lower the temperature, the slower is the reaction. Thus, although the winter may last for more than 6 months, at below 0 degrees Centigrade (32 degrees Fahrenheit) the amount of energy required to sustain the plants will be very small.

The major limitation on these plants appears to relate to the problem of packing their whole life cycle into the few hectic days of high arctic summer. The sedge copes by being a biennial perennial; each new flowering shoot emerges one year then sits tight and completes flowering in the next season. The saxifrage, on the other hand, can sidestep the whole issue by forgoing the luxury of sexual reproduction and producing bulbils in place of its flowers. Bulbils

Above A lemming feeding on tundra grasses. These animals are important links in the arctic food chain.
Right The world's longest distance migrant, the Arctic Tern (*Sterna paradisaea*), nests in northern Alaska and winters in Antarctica, thus enjoying more daylight hours per year than any other animal. Birds flying in from other regions aid in recolonization by bringing in seeds and spores.

The lichens

These dual plants are able to inhabit some of the most inhospitable regions of the world. *Right* Lichen growing on a tree. *Far right* Diagrammatic section showing the structure of lichen.

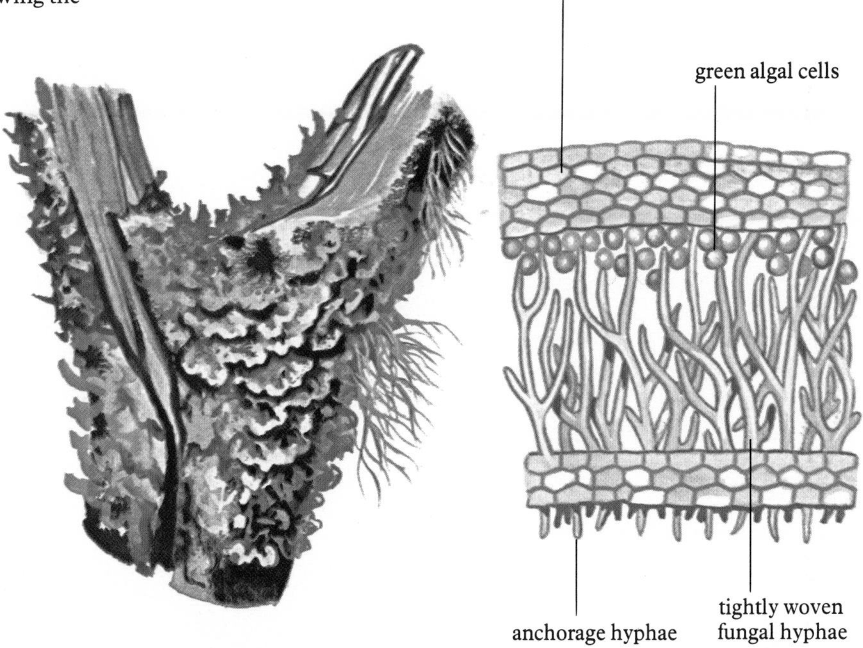

Lichens and mosses – partners on the tundra. The lichens are in 'fruit' and the mosses are in 'flower', the inverted commas indicating that both terms are used very loosely. The real 'fruit' of the lichen is in fact the spore-producing structure of the fungal partner (see above) and the red 'flower' of the moss is a cup of leaves in which the organs of sexual reproduction are borne. Raindrops falling into each cup of leaves splash the sperms out with violence, dispersing them across the tundra to the waiting archegonia.

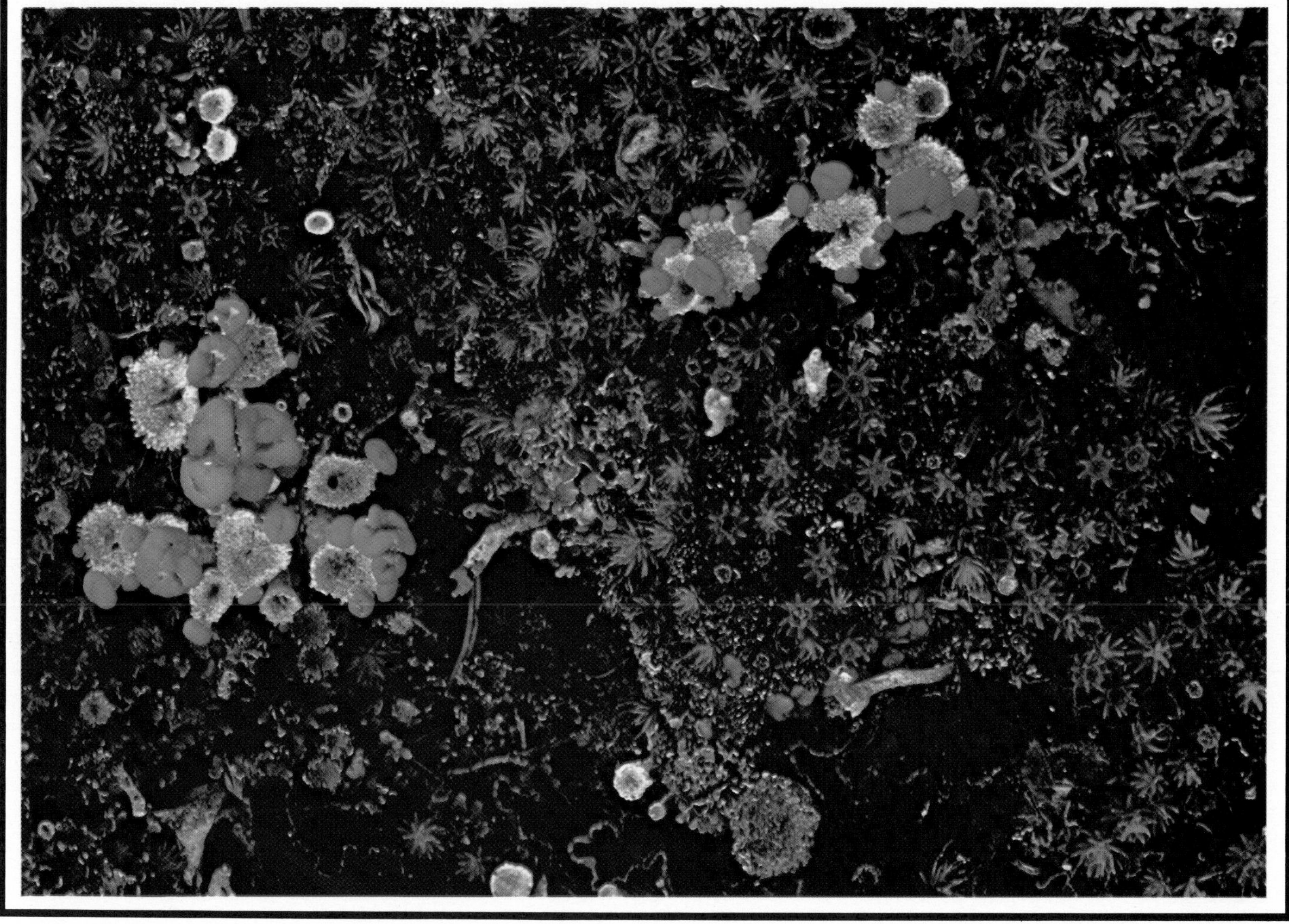

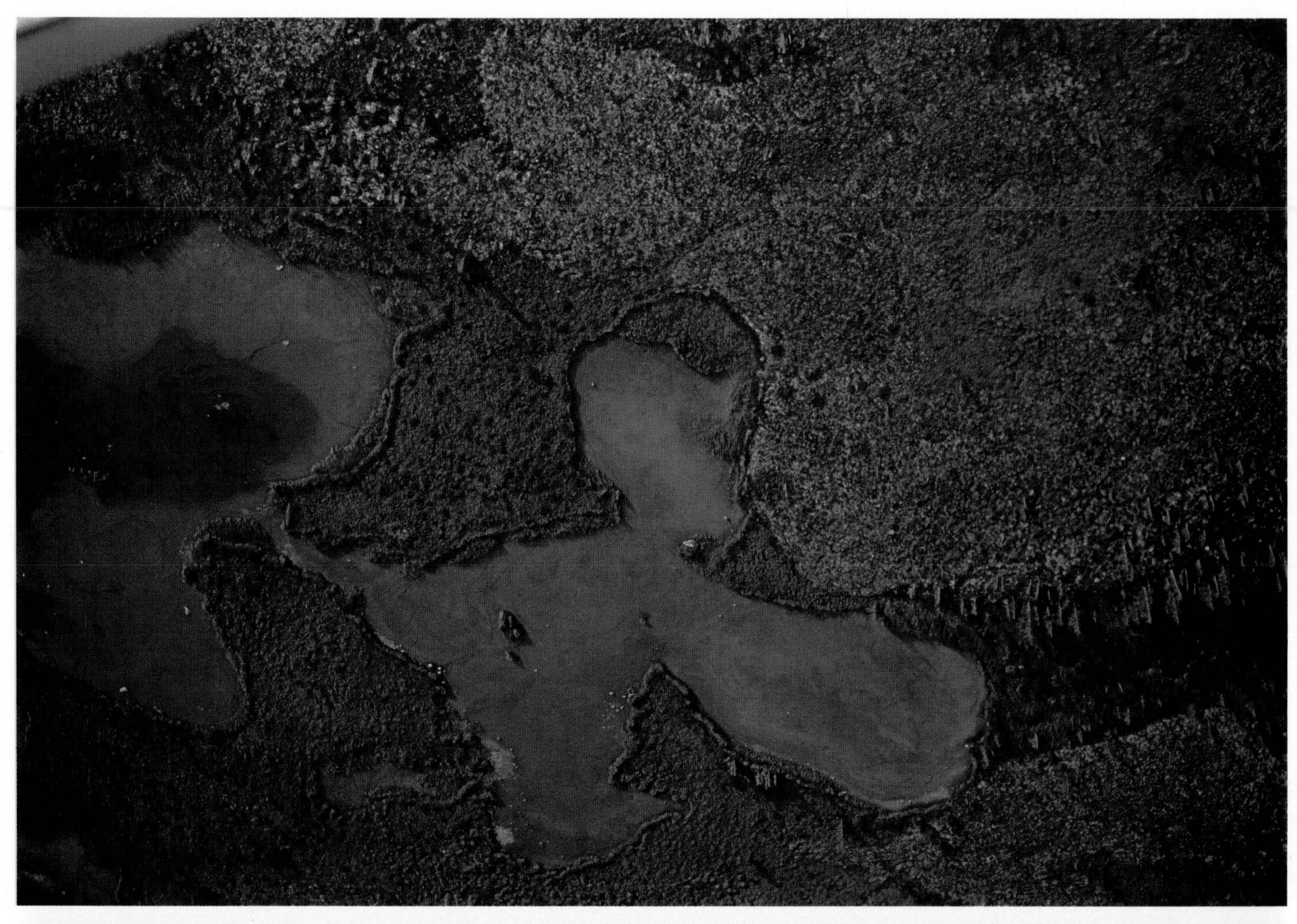

Terrazoid (*top*) and marbloid (*above*), two of Bill Radforth's air-form patterns which allow him to read the potential of each landscape.

Overleaf The edge of the taiga. Stunted conifers, spruce and larch, mark the northern edge of the great belt of coniferous forest that circles the northern hemisphere. Further north the growing gets too tough even for these hardy trees and the instability of the active layer above the permafrost will cause them to topple over. If the climate of the north ameliorates any further, the forest will march on, but if the ice sheets begin to advance, as they well might, the taiga will be forced to retreat once again.

are organs of vegetative reproduction which, when detached from the parent plant, can grow directly into a new one. In this way the saxifrage can fit its life cycle into the length of growing season which any year has to offer.

The dead symmetry of the ice-worked landscape begins to disappear under a new mantle of living green. All that is needed to complete the transformation is time and a continued supply of new seeds and spores. This is, of course, untrue for the really high arctic latitudes, for even if the polar ice cap melted completely it would still require a very large change in climate to make possible the growth of anything but these few hardy plants so close to the pole. The reason is that at very high latitudes the sun shines on the surface of the earth at such a low angle of incidence that its energy is dissipated over a very large area. Hence, unless the sun gets a lot stronger or the earth tilts dramatically on its axis, the high permafrost will never be melted.

Instead of waiting here for the transformation that may never happen, we can travel away from the ice cap south to the warmer lands which have been free from ice for a longer period and which warm up to higher temperatures for longer periods each year. It is thus possible, at least in the mind's eye and with the aid of a helicopter or float plane, to follow the migrating birds and fly like a time machine over the future phases of recolonization. It is the same birds that return in their millions each year bringing with them, both internally and stuck on to their feet and feathers, the seeds and spores of some of the new plants. The birds come to seek out the vast solitude of countless shallow lakes in which they will court, mate and raise the next generation of 'inter-flora high fliers'.

There is one person who has in his time probably done as much tundra hopping as most of the birds. His name is Bill Radforth and he has made it his life's work to try to understand the living landscapes of Canada and especially those areas which are covered with peat. Canadians call their peatlands muskeg and they have got an awful lot of it–129·5 million hectares (320 million acres) is probably a gross underestimate. On a number of occasions, Radforth's work has taken him down from the ice cap towards and beyond the Great Lakes and he has summarized the zones of vegetation and hence the possible phases of recolonization in terms of very descriptive air-form patterns.

The first major zone of vegetation he calls terrazoid because from above it looks like a series of green-brown terraces of peat that are growing out over the dead terrain, obliterating the symmetry of the ice-worked landscapes. The predominance of peat in the high tundra is due in part to the fact that the active layer remains saturated with water throughout the year, which effectively excludes oxygen and hence oxidative decay, and to the very low temperatures which also slow down the decay processes. As the peat builds up to form the plateaux, the permafrost moves up into them forming hidden ice stalagmites, protuberances from the main sheet of ground ice. If the conditions are right, these peat plateaux not only grow sideways but upwards to form mounds which may be over 100 metres (330 feet) long and more than 10 metres (33 feet) high. Such mounds are called palsas and each one has a centre of ice for, like the rest of the terrain, only the surface melts each spring.

The most important peat formers in the world are members of a rather strange family of mosses, the Sphagnaceae, and one of them, *Sphagnum fuscum*, the Brown Bog Moss, plays the major role in building the palsas. On close inspection the bog mosses appear to have all the wrong adaptations for life in a supersaturated habitat. Like all true mosses their leaves are only one cell thick and yet each leaf is made up of two distinct cell types. Narrow, almost worm-like living cells, which contain all the chlorophyll and put the energy into the system, are interspersed among large dead cells. The latter have spiral thickenings on their walls that hold the cell wide open and large pores which let the water in. As each separate plant is made up of many thousands of such leaves and the plants grow very closely packed together, the whole thing acts like a sponge and the plant mass carries its own water supply up with it as it grows. Only the leaves at the top of each stalk get enough light to carry on photosynthesis, the oxygen so released escaping away to the air above. Lower down, in the dead mass of leaves, there is little or no oxygen and hence the bog mosses decay only very slowly, producing the peat.

Thus, the bog mosses and the peat they produce form the substrate on which all the other plants must grow. Many of the members of the associated flora show various adaptations to minimize water loss and so, like the sphagna, they appear to be adapted to a much drier habitat. The wetness of the habitat is in part an illusion brought about by the fact that most ecologists visit the tundra in the height of the summer when the active layer is fully charged with melt water. It is a very different place at other times of the year because the ground water supply is frozen and completely cut off, and any shoot protruding up into the cold arctic wind would tend to lose all its water, unless it were xeromorphic (that is, with adaptations which prevent excessive transpiration).

Modifications range from tiny leaves with thick waxy cuticles, through sunken stomata and rolling leaves that are covered with hair, to the 'throw the problem away' deciduous trees. The first few include many members of the heather and crowberry families and the last, the arctic willows and birches, the gnarled, twisted trunks of which indicate their tree-like nature but their stature is rarely more than 10 centimetres (4 inches) high. Their bonsai nature is partly due to the instability of their habitat, since a large tree would have little chance of standing upright in the midst of all that heaving frost. A second factor is that the best protection against the rigours of the winter, when on a windy day temperatures may fall to an effective −120 degrees Centigrade (−184 degrees Fahrenheit), is to keep a low profile with the shoots close to the ground and, if they are lucky, covered with a nice warm blanket of snow.

In the spring the surface of the peat plateaux is a riot of colour as the tundra flowers come into bloom. The white lint of the willow catkins looks

like the last vestiges of snow and the air is buzzing with the myriad insects which have completed their job of pollinating the willows and turn to other pursuits. Blue lupins, white mountain avens, yellow saxifrages, pink bistorts, silver grey mugworts, the golden succulence of arctic cloudberry, black cranberry, red cowberry and the backdrop of the mosses in all their massed glory–high arctic summer really is a magic experience spoilt for us humans only by the hordes of biting insects. However much you console yourself with the fact that without a meal of blood many of the insects would not be able to complete their life cycle, you can not help thinking that perhaps a few less insects in next year's generation would be a very good thing. Such thoughts are, of course, rubbish because the insects are part of the cycle of life. Fewer insects and certain birds would starve, no birds and many of the carnivores would have to go and, perhaps most important of all, no pollination and the willows would soon disappear, the clocks of succession would slowly turn back and the landscape could become devoid of plants once more.

Further south the lint of the willow has long since blown away and yet there are still splashes of white dappling the brown landscape. These resemble snow, but they are unlikely to be so because we are travelling south towards the warmer climes. As the journey progresses the white expands, blocking out the masses of the peat plateaux which are themselves expanding, devouring all the final vestiges of the ice-worked patterns that helped to initiate the whole process. From high above the terrain looks not unlike a sheet of marble waiting for its final polish, only the streams and rivers of the drainage axes showing through the white sheet. Radforth calls this marbloid terrain. On the ground it is at once obvious that the white is not snow but lichen and yet the glare thrown back from their whiteness is almost as much as that from newly fallen snow. The lichen, which is crisp under your boots, not only helps to feed the Musk Ox (*Ovibos moschatus*) and the caribou (*Rangifer*) but also plays a key role in this phase of succession.

The growth of peat requires a continuous supply of water to keep the whole mass saturated. Since much of the central part of Canada is far from the sea it receives very little rainfall. Nevertheless, large tracts of these dry lands are covered with a blanket of peat and, although I can not speak from experience, it should be possible to walk right across the continent without stepping off the stuff. The answer to what seems an ecological paradox–lots of peat and little rainfall–lies in the permafrost, for much of the water supply of the embryo peats comes from below, not above. This is alright in the high latitudes where the active layer is very shallow and, therefore, remains wet throughout the summer. Further south, the energy of the sun should penetrate deeper, melting the ground ice and so allowing the surface layers to dry out and peat growth to slow and stop. Now we come to the fascinating bit. As the surface layers dry, conditions become perfect for the growth of the large foliose lichens, such as the *Cladonia*, which are so shining white that they reflect much of the sun's energy and protect the permafrost. The active layer, therefore, remains shallow and peat formation can continue.

To prove it all you need is a long sharp stick and/or a bag of soot. With the former you will soon find that beneath the white carpet solid ice is only a few centimetres away, while below the green-brown vegetation of the stream banks it may be more than 2 metres (6·5 feet) down. The soot, liberally sprinkled over the lichens at the beginning of the growing season, causes an isolated area of deep melting and the palsa plateau will probably begin to collapse at that point.

Under natural conditions the palsa plateaux continue growing both outwards and upwards and it is on the margins of these peat mounds that the first trees of any real stature begin to make their appearance. Black spruce and larch are the first, for this is the northern edge of the taiga, that great belt of coniferous forest that spans the globe and marks the next phase in the process.

However, before the taiga really takes over the landscape there is one more major change which is again best viewed from the air. In places the marbloid air-form pattern of the lichen blanket is gradually replaced by a vermiculoid (worm-like) pattern in which sinuous ridges appear to run at right angles to the slope of each area.

Ground inspection reveals that over much of the terrain peat growth has come to a stop and, in places, active erosion of the drier peats is underway. The growth of peat is continuing apace only in the areas which are well supplied with water, and the worm-like pattern shows that these are on shallow slopes which receive run-off and drainage from above. In such areas peat growth continues in the form of ridges called strings, separated by furrows called flarks, which together impede the flow of water downslope like a series of cofferdams and allow the peat to grow. The last vestiges of the permafrost are to be found in the strings for at this latitude the angle of incidence of the sun's rays is great enough and the growing season long enough to melt the vast majority of the ground ice.

The permafrost has played its role preparing the terrain, initiating and supporting the growth of peat. Likewise, the eroding peat now provides sufficient organic matter to speed the process of soil formation and the development of the forests of the taiga.

The vast area of land that for so long lay under the influence of the great white death is nearly ready for the growth of mixed deciduous forest. This is the type of vegetation that above all others makes most use of the opportunities offered by the seasonal climates of the temperate latitudes. In spring the energy stores are mobilized and the sap begins to rise taking with it the promise of a new canopy of leaves. Before this comes to fruition the ground flora must complete much of their life cycle before the living sunshade becomes too effective. Summer is an intermix of glaring sun and deep shade, of sweet fragrance and honeydew, which, together, provide so many animals with their habitat and sustenance. Most beautiful of all is autumn when the leaves having completed their jobs and filled the energy banks of the forest system lose their chlorophyll, allowing a fleeting

Some of the flowering plants which beautify the tundra: *top left* mountain avens; *top right* Siberian Phlox; *above left* Purple Mountain Saxifrage; *above right* Richardson's Saxifrage.

Overleaf The tundra in bloom against a background of eternal ice.

glimpse of all their other colours which have remained hidden throughout the heat of the summer. This is a time of fruits and nuts, the smell of fungi and of ripe decay as the leaves return their minerals to the cycle of the soil from whence they came. A time of hibernation, for the animals of the forest need not migrate to find their winter feed, they simply retire on an affluence of fat to sleep the winter through. Winter is also a time of rejuvenation. The frost bites deep, cleansing and stirring the soil, helping to banish disease and bring new minerals into cycle ready for the spring which is at the most six months away.

In places across the great continent of North America the forest gives way to enormous tracts of grassland, best called prairie or steppe. Today it is generally accepted that the presence of most of the natural grasslands over the world are due, at least in part, to the holding back of the natural process of succession by fire and grazing. However, a look at any vegetation map will make it clear that the grasslands are best developed in the drier areas, and this fact could in part account for their presence. Certainly the boundary between grassland and deciduous forest exists in some sort of dynamic equilibrium and did so long before man came on the scene. In numerous places not too far from the boundary, the unmistakable remains of the burrows of rodents which can only live on the open prairie are found in the soil beneath mature stands of timber. A drought or a severe fire can push the boundary back in the other direction.

In the same way it must be remembered that all the zones of vegetation described above are themselves held in quasi-equilibrium by the interplay of bed-rock, landform, animal migrations and climate. It is sobering to remember that, however warm it is standing in this broad-leaf forest in its summer dress, the same land was probably under ice less than 20 000 years ago. The proof is there in the deposits of peat that bring diversity to the forest floor. In each one part of the history of the process of succession is held in detail. It is not a rare occurrence to find a continuous record that dates back to the first few growing seasons after the melting of the ice; a complete record that has missed none of the important things that happened in the landscape, and that includes the arrival of man.

As the last four periods of glaciation have waxed and waned another important change has been taking place in the water resources of the earth. Whether it should be likened to a new change in state is difficult to say for, although it only pertains to the 0·0001 per cent of living water, its effect is today much more widespread than that of the ice at maximum glaciation. The ascent of man was underway.

When Charles Darwin and Alfred Russell Wallace first proposed their theory of evolution it sparked off one of the greatest controversies of all time, and one that is still going on even today. The argument revolves around the ancestry of man and the poles of opinion are created with the angels, and evolved from the apes.

There is no doubt that, taxonomically speaking, man is a member of the order Primates. We share with the tarsiers, lemurs, marmosets, new and old world monkeys, and great apes certain key features, not the least important of which are four limbs that terminate in five digits and a large brain capacity. The former, though a primitive characteristic, is ideal for grasping objects and the latter for grasping situations. Man is set aside from all other primates by his nakedness, his upright stance, his very large brain capacity, his sophisticated method of communication, his consciousness and his worldwide rise to power.

Recent findings are enabling us to piece together the more solid remains of our immediate ancestors and, although we can only guess at the level of intellect from the dimensions of the fossil crania, it would appear that the first steps in our direction happened about 14 million years ago. Somewhere in the grasslands of the old world a small ape, *Ramapithecus*, left the safety of the trees and the affluence of forest life to seek out a new style of existence. His own ancestors had descended from the trees to take on an upright though stooping stance, a new member of the ecosystem of the savannahs.

What was it that brought about this important shift in habitat preference? We can only guess. The current theories range from competition with more 'swinging' members of the tree-dwelling set, through climatic change that brought seasonality to the productivity of the forest, to more massive changes that favoured the development of grassland. Whatever the cause, there is no doubt that once down on the ground, bipedalism conferred on *Ramapithecus* the advantages of speed in chase and escape, and of being able to look out over the low vegetation.

Some 10–12 million years later the lineage split giving rise in all probability to three lines of descent, one of which has been named *Homo habilis* from his likeness to man and his ability to fashion tools. This, the first man, looked out over the flat open world of rolling grasslands more than 2 million years ago in East Africa, where he was far removed from the effects of the great glaciations that were then underway.

Although many people have and are playing roles in piecing together the story of man's descent, the most famous are members of two generations of a family called Leakey. The latest in this distinguished line of anthropologists is Richard and it is his opinion that it was the need to hunt larger and stronger animals than himself which helped man become a social creature. All members of the genus *Homo* are in physical terms puny when compared with much of their prey and most of their predators. The need to hunt led to a need for mutual help; alone the naked apes would have found survival a real problem, together they could thrive. By 1·6 million years ago, *Homo erectus* was predominant in the line. His brain, having a volume of around 900 cubic centimetres, was well over half that of *Homo sapiens* our own great, great, great . . . grandaddy who exercised his own particular brand of consciousness for the first time only some 100 000 years ago.

What is this thing called consciousness? All living organisms receive stimuli from their environment, and all are irritable in that they can respond to these stimuli. Man responds to stimuli in a reasoned (I do not say reasonable) way. I personally do not think that other animals have that ability but it must be remembered that my opinion may be biased by the fact that I am conscious of being a man.

Consciousness is much more than instinctive or learned behaviour, it allows us to receive a complex of information and, after turning it over in our minds, to come to a series of decisions which may in no way relate to the stimuli or to any previous experience. But what about all the other social animals, do they not have similar abilities?

There is little doubt that honey bees can go out from the hive, search for and find a source of nectar and then return and tell the rest of the members of the particular society more or less exactly where the nectar is. The worker bees could thus be said to be conscious of the needs of the hive and of their role as providers. There is little doubt also that many of the insects which inhabit the top few centimetres of bog moss plants select the habitats in which they feed or rest in relation to the daily march of temperature and humidity. They could, thus, be said to be conscious of their optimum habitat conditions and laboratory experiments have borne out the field observations. Yet, in the same way, a seed that can only germinate under the conditions of fluctuating temperature, which typifies the habitat in which the adult plant will thrive, is selecting its habitat. The fact that the seed can not move of its own free will is of little consequence, for neither can the insects if the conditions are not right around their particular clump of *Sphagnum*. Neither the seed/plant nor the bee/hive system is conscious of the problem. Man is, that is why I am writing this book.

That is not to say that other organisms will never develop our type of consciousness; given time and the 'need' they probably will. However, at this particular state of evolutionary time man is set apart from all others by his own brand of consciousness. There are at this moment some 16 000 million litres (3 520 million gallons) of conscious water on the face of the earth and at least 100 litres (22 gallons) orbiting in space, and it has all happened in the last 100 000 years.

The rise of modern man

The main line of 'advance' to modern man began some 14 million years ago when a small ape, which we now call *Ramapithecus*, walked upright on to the warm dry grasslands of Africa. His direct descendants were doomed to extinction but he gave rise to one successful line of descent – take a look in the mirror and see for yourself how successful it was.

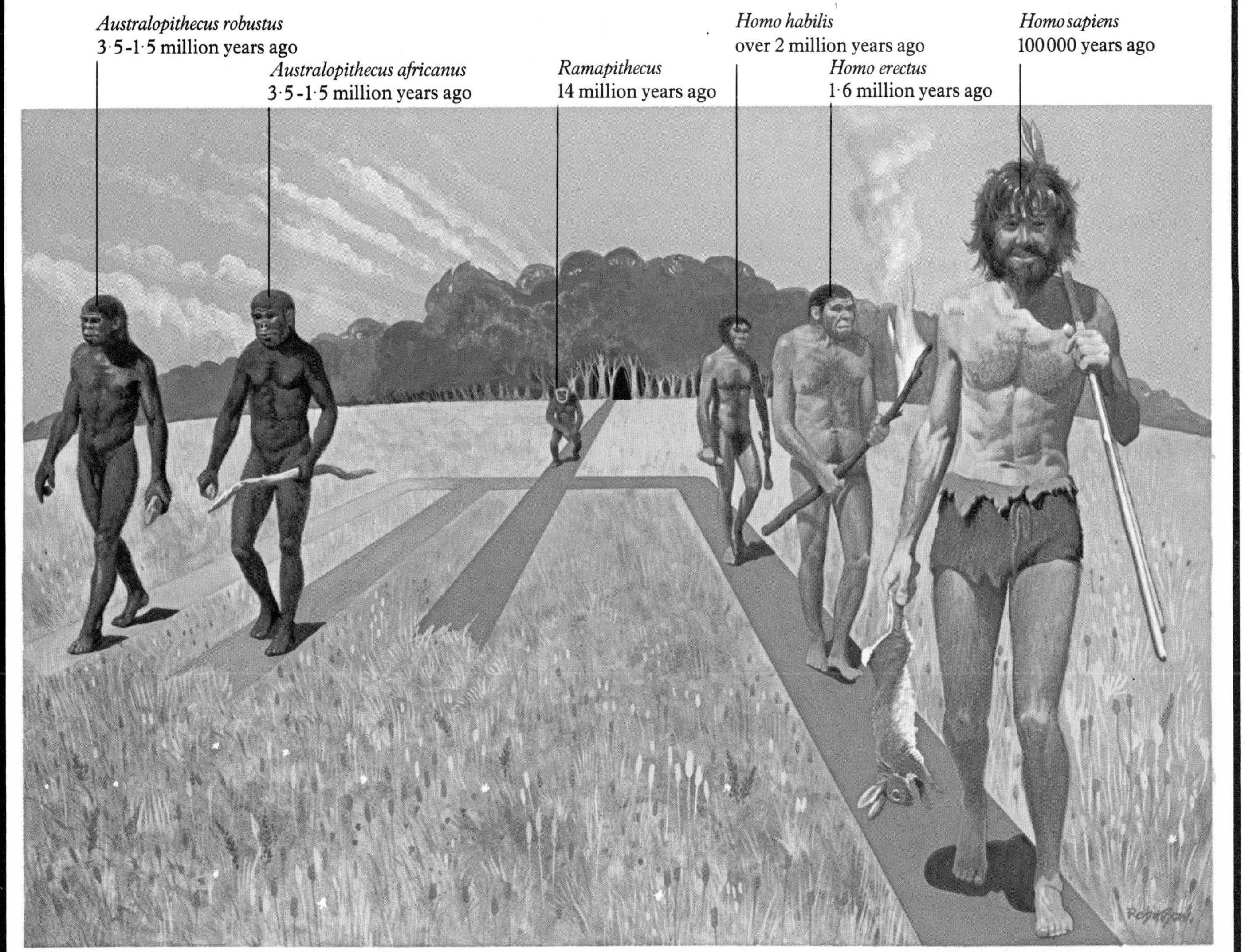

Chapter 8

Life on the limit

Of all the products of evolution, *Homo sapiens* has had the most rapid rise to world power. His success is in no small part due to his evolution of consciousness, an attribute which sets him apart from all the other products of evolution.

Conscious thought has allowed him to side step the slow lane of Darwinian evolution and move into the fast lane of directed advance. The latter is based on knowledge passed on from generation to generation by word of mouth aiding that which is passed on through the normal channels of heredity.

Such unprecedented development on a global scale has not been possible without an enormous strain being put on the support system, the living world on which all the products of evolution, including man, ultimately depend.

There is nowhere better than the subcontinent of India from which to relate this part of the story for here twentieth-century man is still closely rooted in the soil that has fed a large population for the past 2,000 years. The world should not blame India for her mistakes, it should only learn from her experience.

Take a test tube, fill it with a solution that contains all the raw materials that are necessary for the healthy life of a particular sort of plant plankton and stand it in a growth cabinet at the right temperature and in which there is sufficient light. After giving the whole thing enough time to settle down, inoculate the solution with 100 plankton cells, and then sit back and watch what happens. In time the culture solution will turn green and the colour will get deeper and deeper until great lumps of a grey-green substance begin to fall to the bottom, and the liquid becomes clear once more.

If you could keep a running count on the total number of cells present in the test tube and plot them on a graph of cell number against time you would find that the rate of increase of the plankton population goes through three distinct phases. The first is called the lag phase and during this the rate of increase is very very slow. It is now widely accepted that this is due in part to the artificial conditions of such an experiment. Some of the original 100 cells may be too old and some too young for reproduction and hence divisions of the others would have to take place compensating both for the oldsters and the youngsters before a true increase in population would become evident.

From this point on the population passes into its exponential phase of growth–1 cell becomes 2, 2 become 4, 8, 16, 32, 64, 128, and so on. The population doubles at each division, the time taken depending only on the time it takes for each cell to reach reproductive maturity. Starting with 100 cells, they will divide to give 200, 400, 800–well, you work it out.

Plankton cells are minute when compared with even the smallest test tube and so the population can increase enormously before all the available space is filled up. However, long before that happens the population moves into its final phase of senescent growth during which the rate of division slows and eventually comes to a standstill. Plankton do not only require space, they also need raw material for growth and, similarly, they do not only produce new cell material but also waste products which must be disposed of. In the

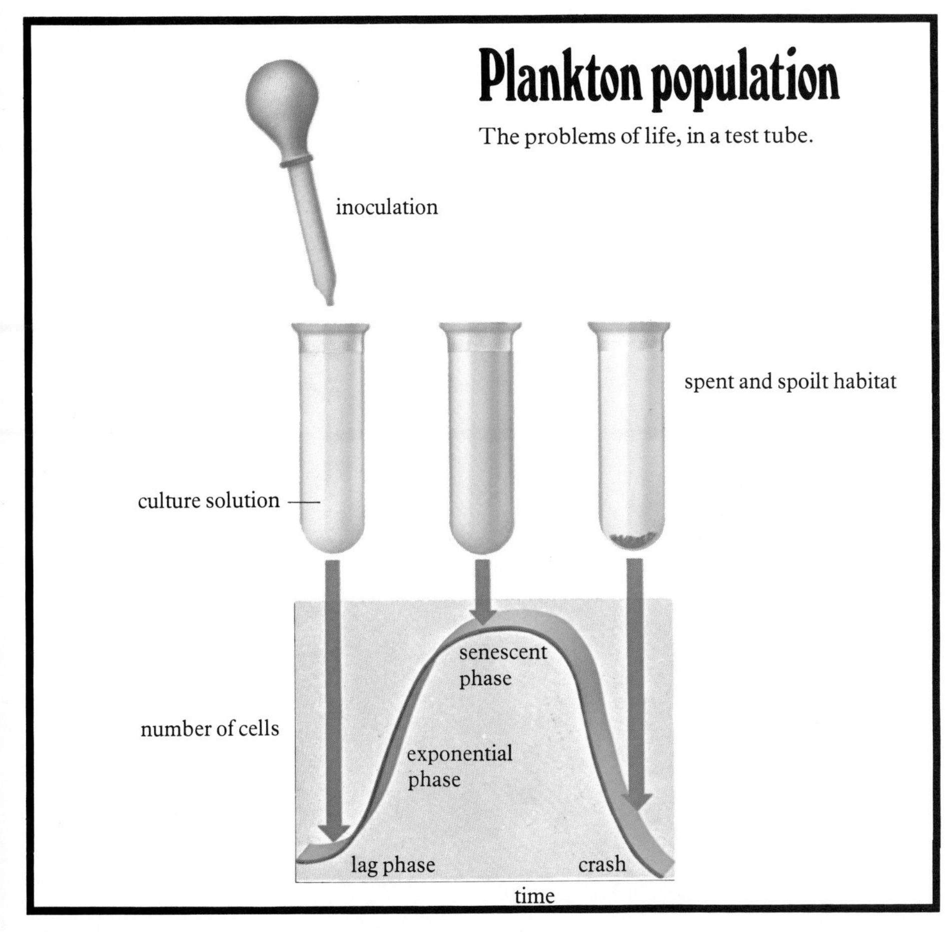

senescent phase the plankton start to use up all the raw materials, they shade each other from the light and taint their environment with toxic waste. The end is near, the population collapses and masses of dead cells sink down to the bottom of the tube.

The same is true for any population be it bacteria or elephants. In any habitat with finite resources their population will follow much the same pattern of development. Under natural conditions there are, of course, many complicating factors; competition with other organisms, direct predation, and instability of the environment will all take their toll and put a lot of wobbles in the graph. Take, for example, the growth of the human population. For 90 000 years *Homo sapiens*, though walking upright, maintained a low profile at least as far as population was concerned. During this immense phase he fashioned stone tools which helped him to gather, catch and kill his food as he spread across the globe. He reached the Western Hemisphere in about 30 000 BC and as he went he opened up for himself the potential of new types of food plants and immense herds of new game. His progress as a hunter is recorded, in all probability, in the extinction of the sabre-toothed tiger (*Smilodon*), woolly mammoth (*Mammuthus*) and giant ground sloth (*Megatherium*) and the tools of his trade litter the fields of the world. During this time he lived as part of the natural living systems, subject to their excesses and limitations.

In about 8 000 BC man stood on the brink of agriculture, his total world population being then no more than five million. This latter fact indicates the stringent limitations under which early man was living and the enormous changes that were to come about with the advent of agriculture. It had taken man more than 90 000 years to reach his first five million, but with agricultural produce firmly under his belt it was only going to take another 1 500 years to double this figure. What was it that made man take this most significant step, putting aside the 'free' life of a nomad to take up a more settled existence based on husbandry? Was it simply the culmination of a gradual increase in conscious knowledge, passed on from generation to generation, or was the transition brought about by necessity?

Evidence suggests that some of the first steps towards the agricultural revolution took place somewhere in the fertile crescent along the hills that now form the borders between Israel, Jordan, Turkey, Syria, Iran and Iraq. The potential was there, at least in comparison with the adjacent areas of semi-desert. One school of anthropology believes that nothing more was required–the potential was there

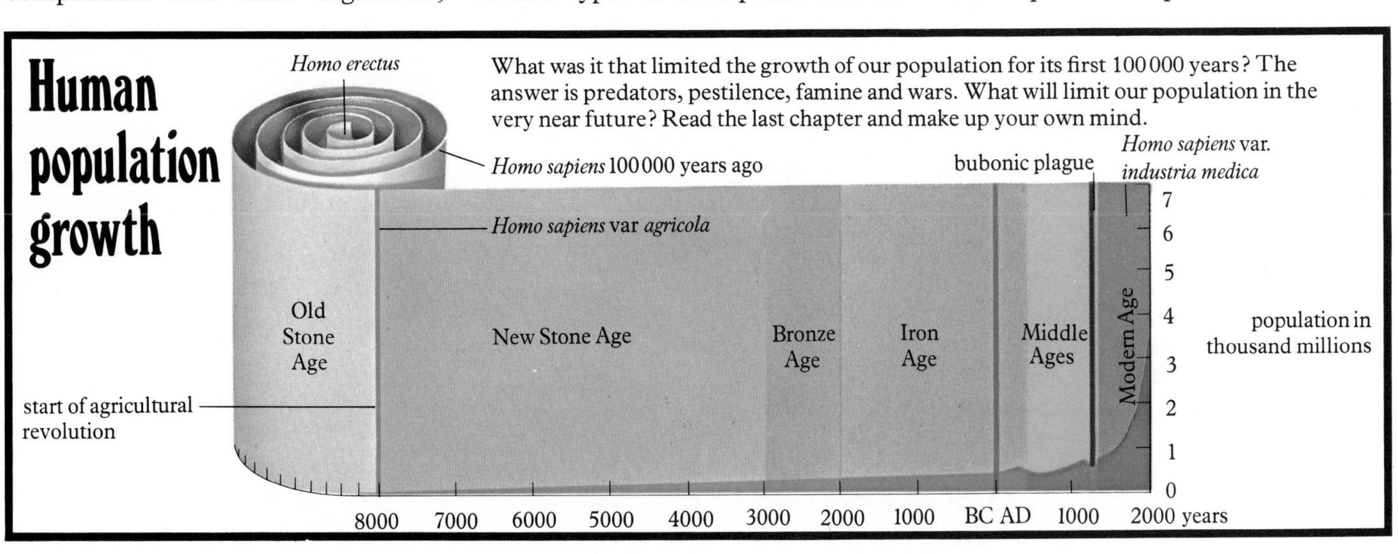

and man was ready to use it, so it had to happen. The other school looks to the areas of semi-desert for the answer. Their argument goes something like this. As the increasing population of hunter gatherers began to overstress the natural productivity of the most fertile areas some of them would have been forced out on to pastures less green. Faced with the problems of life in the semi-desert, their attention would have been focused on the more fertile spots around oases. There they would have stayed, perhaps long enough to get in tune with the local environment, learning the natural cycle of events, what useful plants grew where, and when and which of the animals that dropped in for a drink could be most readily captured. With such knowledge the cultivation of crops and the husbanding of animals would have been very easy.

Was it an accident or simply his increased knowledge that made the first farmer realise that the grains he usually ate could be planted to produce more grain-bearing plants? It is easy to imagine a fortunate accident when an unexpected period of prolonged rain caused the contents of a grain store to begin to germinate. Equally, an oasis dotted with the shade of palm trees and a ready supply of food on tap would be the ideal place to just sit and think it all out. Did he fall or was he pushed, accident or design? However it happened, his consciousness allowed *Homo sapiens* to sieze on the opportunity.

It was certainly no accident that the start of one agricultural revolution took place in the fertile crescent for this was the native home of both *Hordeum spontaneum* (wild barley) and *Triticum vulgare* (wild wheat), the two grasses that were to form the staple of the new farmer's diets. There are two problems in using the seeds of grasses to form the basis of your diet. First, the grasses did not evolve nutrient rich seeds for the benefit of passing herbivores, but for the procreation of their own kind. For this reason the fruits of wild cereals are readily detached from the plant and when ripe the slightest shake sends them all over the place. Now that is ideal for the plant because it disperses the seed, but think of the poor farmer. It is, therefore, very likely that, right from the start, he would have selected his seed mainly from the plants that did not shatter too easily–a real case of unnatural selection. The second problem is one of food preparation. Unless you like a lot of roughage the cereals must be threshed and winnowed, and the fruits ground into flour. All that work requires not only a lot of energy but also some pretty sophisticated hardware including milling and pounding stones. Was it the hardware that became the millstone around the neck of nomadism and finally fixed man to his farming spot?

A group of hunters living in a forest might clear an area around their camp in order to gain protection from surprise attack either by enemy or predator. New plants would spring up in the clearing, some of which would be of direct use to the family and others would attract game into

A market near an oasis. What better place to meet and talk over an agricultural revolution? Oases such as this have always formed a focal point in the life of nomads who inhabit areas of semi-desert.

the clearing, especially during the winter. The seeds of animal husbandry could thus have been sown, the group discovering the animals most amenable to domestication—dogs for hunting, goats, sheep, cattle, pigs and horses for meat and hide. The clearings would thus have been extended and dwellings made more permanent. The transition from cave to village was underway.

Perhaps one of the most outstanding and important attributes of man is his ability to pass on information. Even in the earliest village site so far excavated, at Jarmo in the Zagros mountains in Iraq, there is evidence of long distance 'trading', for some of the hunting implements found are tipped with obsidian, the nearest source of which is at Lake Van more than 100 kilometres (62 miles) away. If volcanic glass can travel, so can ideas, and when one group is on to a good thing, copying becomes patently obvious. There is no doubt that the new innovations spread like wildfire across the world of man. By 5000 BC farming had spread to the fertile valleys of the Tigris and Euphrates where the silts borne down by annual floods ensured productivity year after year.

Over much of Europe the transition from hunting to farming appears to have been very rapid, suggesting that it was sparked off by the immigration of new ideas or even that the ideas went with the immigrants who were still crowding north into the wealth of the post-glacial spring. Whether the agricultural revolutions which took place not all that much later both in the Far East and in the New World were linked by the flow of ideas is still a matter of conjecture. It seems unlikely that the developments in America, which were based

The fertile crescent

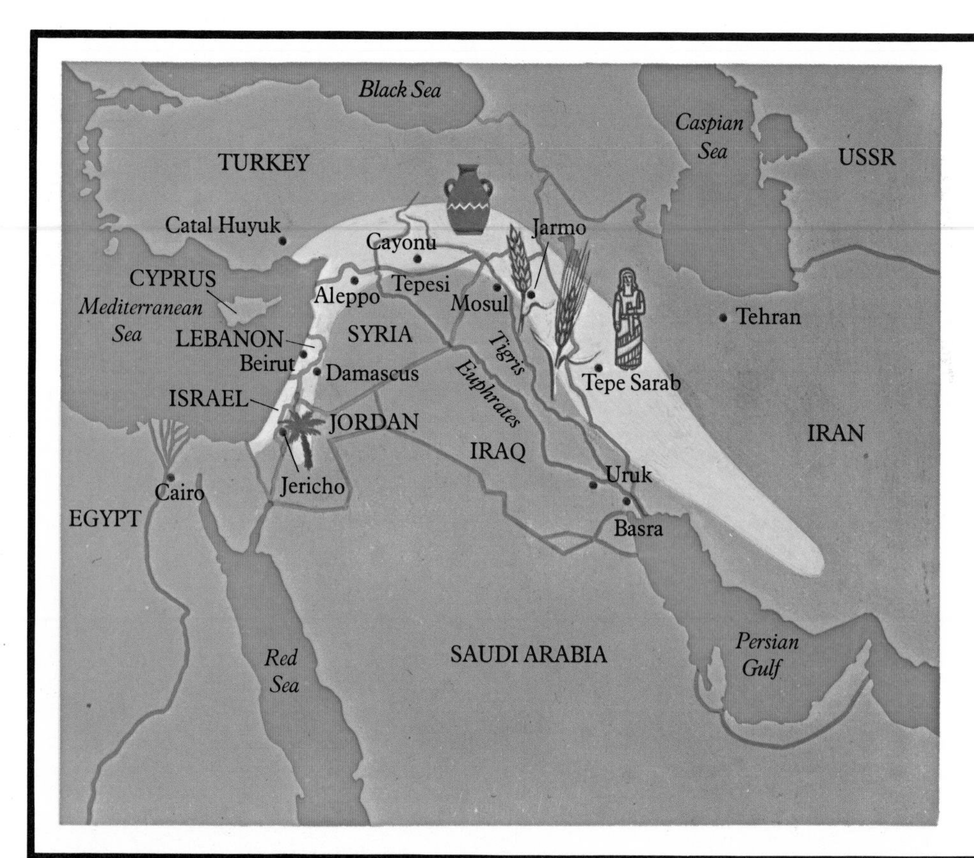

For a long time it was thought that this area of fertile land held the secrets of the first agriculturalists. Although this is no longer thought to be the case, the extensive and detailed work in this area has given us a great insight into the early days of farming. On the evidence to hand more animals were domesticated here than in any other region.

The nine centres of revolution

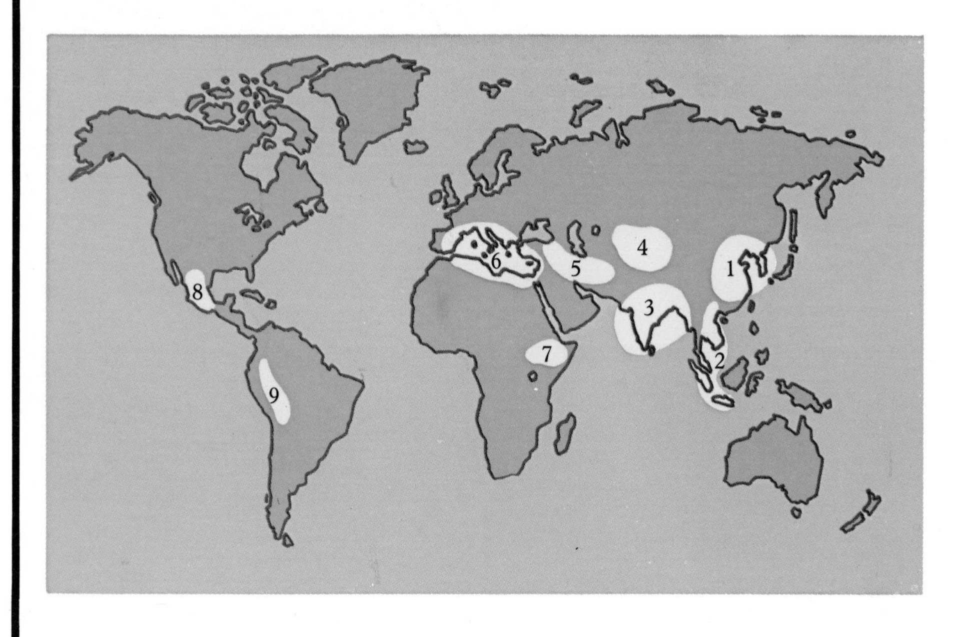

A Russian geneticist by the name of Vavilov believes that these nine areas were the centres from which our cultivated plants were first selected. Centres 1,3,4,5 and 6 produce all the cereals except sorghum which came from Africa (7) and maize which came from tropical America (8 and 9). Where did it all start? The evidence which would enable this question to be answered is still being gleaned from the excavations of the world. At present the indications are that rice was first cultivated more than 15 000 years BP in south-east Asia, but I think that at the moment it is only safe to say that the answer is still in the cooking pots yet to be unearthed.

on an entirely different set of plants and animals, could have been in any way connected. This fact certainly leads one to argue that man, on a world scale, had reached a pertinent level of consciousness at about the same time.

Whatever the answer, once the first faltering steps had been taken, agriculture so increased the potential of the world to man that his population began to shift out of the bottom gear of what appears to be a prolonged lag phase into exponential growth. The strict analogy between the growth of the human population and that of our experimental colony of plankton begins to break down at this point. The development of agriculture by man is akin to the plankton evolving into another form which can more efficiently use the resources of its test-tube habitat. Perhaps from this stage onwards man deserves a new name, *Homo sapiens* var *agricola*.

This is no unique happening. Looking back through the whole of evolutionary time there are a number of milestones, each of which mark a change of pace in the whole process,

3 400 million years BP–The first organism. Organization isolating the living chemicals from the environment.
2 000 million years BP–Photosynthesis and nitrogen fixation. The energy of the sun, free oxygen and adequate supplies of nitrogen became available.
1 500 million years BP–First eukaryotic cells. Further internal organization makes way for the development of many-celled plants.
600 million years BP–Many-celled animals. Larger animals can move more efficiently exploiting the total resources of the sea.
400 million years BP–First land plants. The potential of the other two-sevenths of the earth is up for grabs; the first true soils begin to form.
0·1 million years BP–*Homo sapiens*. Consciousness personified.
0·01 million years BP–*Homo sapiens* var *agricola*.

sparked off by an important evolutionary advance.

It is important to note that each step is cumulative in that it appears to accelerate the process of evolution on towards the next stage. Up to *Homo sapiens* it is easy to understand the cause of the acceleration in terms of a tangible increase in potential, but what about man? Why has his rise to world dominance been so rapid? What is the importance of consciousness to the process of evolution?

It was Darwin who first made the thinking world doubt the dogma of special creation, replacing it by the dogma of the creative power of natural selection. In the many arguments which followed the radical change of attitude to our own origins, one very pertinent factor was often overlooked. That factor was time. God created the world in 7 days, stages, aeons, it does not really matter which because at least that statement recognizes the importance of time and the fact that the creator determined the time it took. In essence, Darwin replaced the absolute of creation by a time-dependent process called evolution by natural selection, a process in which there was no creator to limit the time taken.

The name of the game does not really matter but the mechanism and the time it takes does. Natural selection works against the background of variation in a particular population of plants or animals. In any particular situation where the population comes under stress, those members of the population which have traits (adaptations) which are an advantage in that situation would be likely to produce more offspring than the other members of the population that were less well able to cope with the stress. If those traits could be inherited (that is, passed on to their offspring) then they would become more and more widespread in future generations. Perhaps the best way to summarize it would be evolution by means of random trials with the perpetuation of successful traits.

The problem with such a mechanism is that the length of time that will be required for change could be enormous; and it must be remembered that time will also be a necessary function of the process by which the all-important variation is built into the population. Has there been sufficient time in the last 3400 million years?

Another theory of evolution, Lamarck's theory of the inheritance of acquired characteristics, suggests another possible source of variation which would require less time. If his now infamous population of uniformly short-necked animals, stretching up for the leaves on the trees, could elongate their necks and pass the self-inflicted 'trait' on to their offspring then the archetypal giraffe would not be long away. It was too easy in the light of the developing science of genetics (the science that deals with inheritance) to pour scorn on Lamarck's theory. However, this still leaves the need for a theory to explain how the variation in the inheritable characteristics that are the stuff of natural selection could have come about in the timetable of evolution.

Recent research and thinking is beginning to find answers and is forging links between what have to date appeared to be opposing lines of thought, Darwinian and Lamarckian; and nowhere is there better evidence than when considering the social evolution of man. Through his consciousness and ability to communicate complex ideas man ensured that knowledge gained by one generation was passed on to the next. Thus, at least in social terms, man's evolution was lifted out of the slow stream of 'Darwinian' advance and accelerated into an unprecedented phase of 'Lamarckian' progress. *Homo sapiens* was the first organism which was not subject to the slow process of change based on natural selection, and as var *agricola* he was no longer subject to the limitations of natural productivity, at least in the short term.

We have seen in the earlier chapters that the natural productivity of any area depends on the climate, especially the annual march of temperature and rainfall, the soil and the state of play of evolution at all levels in respect to that environmental complex. The removal of mature vegetation, at least on a small scale, is unlikely to affect either the climate or the state of play of evolution, but it must have an immediate and massive effect on the soil, especially on the supply of minerals for the new 'take-away' crops. The continued productivity of mature vegetation depends

Overleaf A family of farmers at work in the Trisuli Valley, Nepal. There are many jobs to do down on the farm and each member of the family learns to do all of them. There are no free loaders and each unit of the society produces little more than its own requirements, with some to store away for more frugal times.

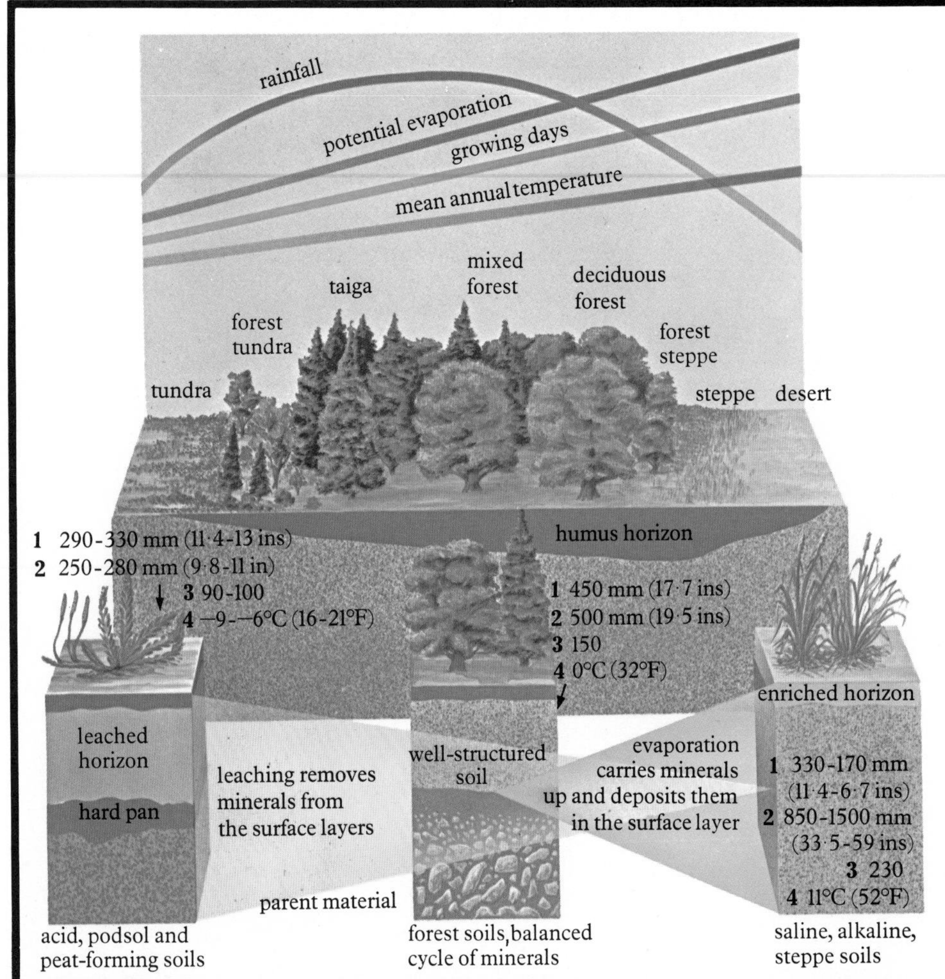

Vegetation, climate and soil

If you look at a map of the world it is easy to see that Russia owns a very large slice of the soils of the world. No wonder then that we are indebted to her soil scientists for much of our understanding of the complex interactions between vegetation, climate and the living soil. The diagram shows a schematic representation of the more important features that link climate, vegetation and soil along a transect from arctic to dry tropical Asia. In essence, warmth and an adequate distribution of rain give rise to the most productive natural soils. The vegetation not only helps to make the soil by aiding the process of weathering and adding the all-important humus which feeds the decomposers and binds minerals and water into the soil giving it structure, it also protects it from the extremes of climatic fluctuation, especially leaching and evaporation.

KEY

1 rainfall
2 potential evaporation
3 growing days
4 mean annual temperature

on a constant recycling of minerals through a great diversity of organisms, and the fact that the recycling process is to a large extent protected from the direct effects of rain and wind by the presence of the vegetation. A crop is usually a monoculture which only protects the soil from the weather over a short part of the year. At the end of each growing season, part of the crop is removed from the field in which it has grown and with it goes some of the nutrient store.

At first the nutrient gap was filled by shifting agriculture. An area that had been cleared of forest was cropped until the natural supplies of minerals were depleted when the plot was left to regenerate and the farmers moved on to turn other forests into pastures new. Such a state of affairs could only continue until the demands of the population of farmers caught up with the production of the available land. Agricultural man responded with all sorts of innovations to improve his plot. Mountain slopes were terraced, dry lands irrigated, and the discovery of the beneficial effects of lime, manure and the growth of legumes on the nutrient store of the soil led to mixed farming and crop rotation. The original selection of the best crop plants by the earliest farmers was continued and eventually supplemented by experiments aimed at breeding new strains and, perhaps most importantly, the right strain for the right environment. Perhaps most significant were the massive migrations that took place, the pioneer agriculturalists following close on the heels of the explorers, taking their knowledge with them to reap the rich rewards of new harvests.

Apart from providing mankind with a bigger and better source of food, agriculture began to release man from the drudgery of a hand-to-mouth existence in which all the hours of the day must be spent in supporting the energy requirements of immediate dependants. Farming meant, at least for some, the promise of spare time in which both hand and mind could be turned to things other than subsistence. The excavations of pre-agricultural sites reveal little other than the remains of the tools of the hunting and gathering trade. Even the earliest agricultural sites are typified by figurines and ornamented utensils, and yet in these we find that each dwelling has a corresponding field system, indicating that every family played their part in farming. Only later do we find villages in which the number of dwellings exceeds the number of farmed units, and it is in these that many decorative religious and other artefacts are discovered.

The societies which produced the pyramids, the Great Wall, the great mosques, cathedrals and churches, the old masters and the great composers were not long in coming, and all were the products of agrarian societies. Man at last had time on his dextrous, five-fingered hands, and with a brain capacity of 1 350 cubic centimetres, nothing was impossible.

Ten thousand years ago there were some five million men on earth. By 1750 AD the number was approaching 1 000 million and only 200 years later the 4 000 million mark was well in sight. If the estimated figures are plotted against time, both on a loga-

Above Valley of plenty in a man-made desert – the flood plain of the River Indus near Leh, India. Wherever agriculture had its real beginnings there is ample evidence that it spread with the flood waters throughout the major rivers of the world. The reasons are simple – an adequate supply of water on tap, a fresh supply of minerals each year on flood and a good local freeway for commercial transport.

Overleaf Brick making the original way. In many cases, the transition from cave to hut went hand in hand with the removal of the forest, part of which was used to build huts and houses as protection from the elements. Agriculture almost 'sprang' up on the cleared ground and as more clearance removed all the trees, new technologies of building and construction had to evolve. Use of the earth and clay on which the trees once grew, mixed with straw from the cereal crops to make bricks seems, at least in hindsight, a very logical thing to do.

rithmic scale, the data reveals three main surges of population increase. The first started more than 1 million years ago and corresponds to the rise of consciousness (the cultural revolution); the second starts around 10 000 years BP and corresponds to the dawn of the agricultural revolution; while the third, which blew the top of the exponential phase of increase turning it into an explosion, corresponds to the dawn of the industrial medical revolution.

If one requires an explanation, a cause for the latest breakthrough, then one can do no better than look through recorded history back to the fourteenth century when bubonic plague wiped out 25 per cent of the population of Europe. This is the only happening that put a significant dent in our population curve and perhaps most significant of all is the fact that it happened in Europe where the bulk of the cultural aspirations of man were then centred.

It is well to remember that an increase in any population is the result of the interaction of three processes, birth rate, death rate and immigra-

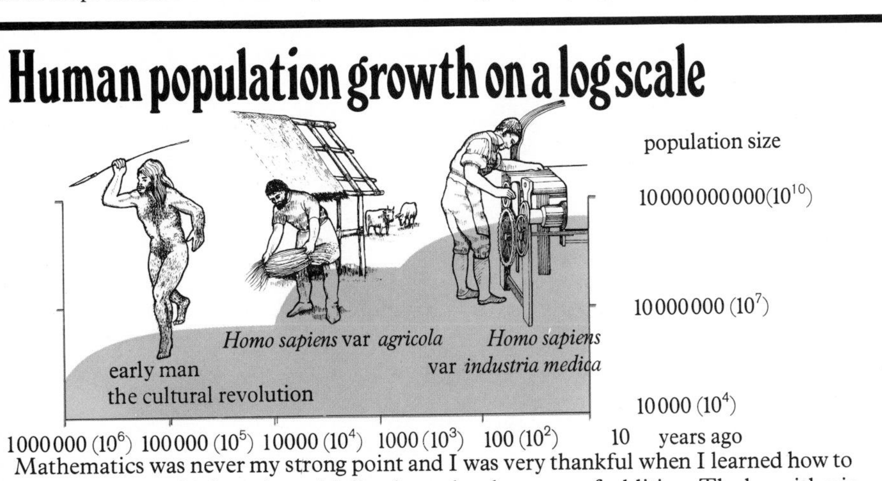

Mathematics was never my strong point and I was very thankful when I learned how to use logarithms which made multiplication a simple matter of addition. The logarithmic representation of the increase in human population over the last million years certainly throws new light on what has happened for it shows that it has taken place in three distinct surges, each of which corresponds to a major change in our ancestors' way of life. (From 'The Human Population' by Edward S. Deevey, Jr.

tion. As migration to or from the world has to date played a very insignificant role, man's population is dependent on the other two factors. Let us imagine that with the world population standing at five million it took off on a phase of truly exponential growth. The important characteristics of the model are an equal number of males and females, each of which reach reproductive maturity at fifteen and die at twenty-five having produced an average of four viable children per couple, all of whom live to reproduce in the same way. If there were no limitations the population would double in less than 20 years and in only 200 would have topped the 5 000 million mark.

The fact that it has taken at least 10 000 years to approach that figure is indicative of the stringencies of famine, pestilence and war that have limited it throughout that time. With the next doubling time computed to be 35 years it shows that the so-called population explosion is in reality only the beginning of exponential growth.

The black death could well have been the cultural shock that spurred thinking men to put their houses in order, sowing the seeds of both the medical and the industrial revolution. The harnessing of steam put massive power in man's hands, furrows could be dug deeper, mines driven faster and the products of the new affluence transported far and wide. Machines may have devalued the might of human muscle but they put more resources within his grasp and provided him with more time to sit and think. The printing press had already taken communication into a whole new dimension and had helped man turn his attention more to his own physical and moral welfare. Hence, advances lifted the industrial world towards the euphoria of exponential increase and as the demand for raw materials, food, guano, and so on, spread across the world, so too did the benefits of this new society.

The main thrust of this development took place during the reign of one queen and much of it took place within the bounds of her great empire. Her name was Victoria and it was her subjects who lifted Darwin's book, *On The Origin of Species*, up to the rank of best seller. One of the main reasons for its immense success was its sub-title, which the author disliked, but which was insisted on by a shrewd publisher. It read 'the preservation of favoured races in the struggle for existence' and it appealed because the Victorians 'knew' they were the favoured race and were willing to swallow the bitter pill of descent from the apes in the knowledge that they would ascend to the angels.

Despite all their trappings of success, the Victorians became famed for their cautionary tales. This seems an appropriate point to sound a note of caution and what better place to do it than in Imperial India.

Cherrapunji, on the Shillong Plateau of Assam, is reputedly the wettest place in the world, receiving, on average, an incredible 11 metres (430 inches) of rain per year. Lying as it does just outside the tropics and enjoying a high level of rainfall the year round, it has an almost perfect non-seasonal climate, with every day a growing day. Altitude brings the main limitations and the natural vegetation is best described as upland, cool, tropical rain forest. It is a paradise both for the botanist and the entomologist, being the home of an enormous variety of plants and insects. The former include a great number of orchids and the latter, the giant moon moths.

The tree layer is dominated by oaks and rhododendrons, which are festooned with liana-like vines and the whole smothered in epiphytes. The overall dampness of the climate manifests itself in great cushions of sphagnum moss that almost smother the trunks of the trees, and the presence of one of the most revolting forms of forest life, land leeches.

We know that man arrived on the forest scene and, despite the leeches, he must have found it a land flowing with all the things a hunter gatherer could desire. During their social

A product of the industrial revolution – lots of energy in a steel works. How much energy do you consume every day? If you live in the western technological world the answer is a great deal, however good you are at sticking to your diet – food on your plate, petrol in the car, and all the energy that went into winning and moving the raw materials, and in making and transporting all the manufactured goods, on which you depend.

evolution as part of the forest system, the local indians developed a very special conscious relationship with the forest, which centred on the worship of the trees. The dogma of their religion was that the trees are the dwellings of the gods, and to damage a tree was, thus, an act of sacrilege, punishable by death. Whatever the rights and wrongs of such a belief and practice, it protected at least some of the forest system from which the indians derived a more than adequate living.

It is of great interest that the influx of new ideas which led to the expansion of shifting agriculture in the area

Above *Actias selene*, a moon moth fit to grace the sacred groves of Assam. Butterflies during the day and moths during the night, this is one rule of the insects the world over, although it is never strictly adhered to by either group. In this way the food potential of an area is shared out, the day fliers feeding and pollinating the day flowers, and the moths coming in on the night shift to get their food from those flowers which produce some of the richest scents of all.
Below One entrance to the sacred grove of Mawphlang near Cherrapunji. The tree rhododendrons are in full flower against a backdrop of the oak *Quercus griffithsii*.
Below right With no trees to hand, animal dung is the only source of fuel to cook the food. The humus and the nutrients it contains will go up in smoke, the cycle is broken and the end is very near. This is one region where alternative technology such as small solar-powered ovens could be of great benefit to the whole system.

must have gone hand in hand with a change in these beliefs. It was probably not the first, and certainly not the last, time that a change to a new materialism led to a change of faith. The effect was catastrophic. The soil, no longer protected by the thick canopy of broad leaves, was open to the leaching and erosive power of the rain. The thick forest soils soon disappeared, colouring the waters of the many rivers that drain the main plateau. For a few years, each forest plot yielded a rich reward of rice and other staple foods, but very soon the nutrient stores were gone and the farmers had to move on.

In the better and least denuded sites regeneration went on apace, the fallow fields being replaced by new stands of tree. But in many places the broad-leaved trees did not return; their place was taken by a conifer *Pinus khashya*, whose tough, needle-like leaves led to the build up of a layer of raw acid humus in the regenerating soils. The acids from this humus aided the continuing process of leaching which carried the nutrients down the soil profile out of reach of even the deepest-rooted plants. Whether, given time, the nutrient stores would have been replenished and the broad-leaved forests would have returned we do not know, because time was not on their side. The enlarging population of new farmers returned to remove the conifers and till the soil once more. If they were lucky some crops were raised, but the end was very near. The ground water could no longer find free drainage through the altered soils and so provided ideal conditions for the growth of the bog mosses and other peat-producing

Overleaf New rice growing on old terraces. Once no more than a wild grass growing in the swamplands of south-east Asia, today *Oryza sativa* feeds two-thirds of the world's human population. One of the earliest developments in farming techniques was to terrace the hills.

A paddy field that never was, thanks to *Eichhornia crassipes.* The field is situated within the campus of the University of Gauhati in Assam. Enriched with organic matter it made an ideal habitat for the spread of the Water Hyacinth. Today work being carried out under the

direction of Dr Baruah in that university is not only helping to solve the problem but is already turning the crop of Water Hyacinth to good use. What was a noxious weed is now fed to cattle. Wherever there is potential it will be used.

plants. A diverse, stable, productive forest had been replaced by a wet desert of little use to man, and all because one member of that forest system had changed his evolved pattern of life.

Like all good Victorian cautionary tales it has been simplified, and the points that the storyteller wants to make have been highlighted. The evidence, however, is there in the abandoned fields now covered with a thin crust of peat; the sacred groves, minute scraps of forest still held in some reverence by the locals; and a population depleted and in part dispersed to the promise of other materialism.

All those nutrients must have gone somewhere, and they did; down towards the great rivers Ganges and Brahmaputra which skirt the plateau at that point. There, on the rich silts brought down by the rivers, agricultural man had for centuries raised a good crop of rice, the cereal that is the staple for two-thirds of the modern world. The paddy fields that lay in the basins of these rivers draining down from the Shillong Plateau must have felt the beneficial effects of all those eroded minerals, but here there is a man-made sting in the tale, and it centres on two forms of pollution. One of these may not even come within your definition of pollution; the other, you will all agree, we can not get along without, for everyone of us does our fair share each day.

One of my favourite sights in the world is a fallow paddy field, for at the right time of year they are a riot of water plants, pondweeds, waterlilies, lotus, rushes, sedges, ferns, mosses, liverworts and algae–a paradise for the aquatic botanist.

Unfortunately, the love of beautiful water plants prompted man to carry his favourites with him as he travelled the world. One such was *Eichhornia crassipes*, the water hyacinth, a native of the river Amazon where it effectively covers and blocks backwaters and lakes. *Eichhornia* was taken to Assam as an ornamental plant and, finding the paddy fields an ideal habitat, it has in places made rice production very difficult. The reason for its wild success in Assam and in many other parts of the world is a combination of its buoyant growth, its rapid vegetative repro-

duction by means of offsets and the fact that in its new territories there are no large herbivores to eat it and no plant with a similar growth form to compete for the niche. The potential was there and *Eichhornia* has filled it in a very obvious way.

During a brief study of the water hyacinth problem in fallow paddy fields around the town of Gauhati in Assam, it became obvious that the distribution of the plant was correlated with the enrichment of the paddy waters by sewage. Close to the centre of the towns and villages the fields were filled with colonies of large 'aggressive' plants, while further out in the country, isolated colonies were found only close to local privies and even beside bus stops and the like where the human presence was greatest. Undoubtedly pollution by sewage greatly aids the pollution of the natural vegetation by the weed *Eichhornia*, at least around the area of my study. Could it be that the nutrients eroded down from the wettest places in the world could have played a similar role, enhancing the spread of the plant throughout the area? I do not know the answer but perhaps it does not matter because the point of this particular tale is to indicate the complexities of problems initiated by man altering the natural balance of any part of the earth.

My third and final cautionary tale takes us way across India to a much drier region of seasonal monsoon

Not so many years ago this was a forest – a caption which fits so many landscapes in man's world.

rains. The exact location is of no consequence because what I am about to describe occurs commonly across the semi-arid world.

Records tell us that this landscape used to be covered with summer deciduous tropical forest, its trees being productive only during the rainy season and aestivating for the rest of the year. Primitive agriculture followed much the usual pattern; shifting agriculture removed the trees opening up the mineral store to 'take-away' crop systems based on upland rice. The mineral store was thus depleted and, although it took a much greater time than in the more extreme climate of the Shillong Plateau, the end point was much the same, a desert of little use to man. However, it is a very different sort of desert, for here the leaching of the rainy seasons alternates with long periods when the soil is completely bared by drought and hence open to the evaporative power of the sun. Water, moving back up through the soil profile, brings with it soluble minerals, some of which are of use to the crop but the vast bulk being of no use at all. The former are in part removed with the crop, the latter stay put and the soil becomes more alkaline and more saline, both of which are conditions which few plants can tolerate. Few plants leads to a depletion of the organic matter present in the soil and, as it is the humus that helps to bind the mineral particles together giving structure to the soil, the structure begins to break down and the mineral particles pack together. The resultant usar soils are so compacted that it is very difficult for the bullock-powered plough, let alone the plant root, to penetrate far down. In the same way, much of the monsoon waters are lost by run-off before they can percolate down into the soil.

The result was the total collapse of the farm systems and the emigration of the farmers, at least those fortunate enough to have a place to go. With man-made deserts so widespread, the spectres of famine and pestilence have ridden over much of India in the past few centuries; so much so that the sub-continent has in recent times become the symbol of the problems of the Third World.

It must, however, be remembered that the lands of India have supported a large population of agricultural man for more than 2000 years. Not a bad record when compared with the farmlands of Europe, and especially of North America, some of which were first put to plough less than 100 years ago. I wonder what the big wheat country will be like in the year 4000 AD? Can we be complacent?

The continued upward sweep of our population curve hides more than it records. In that slow rise are hidden not only the joys of success but the grief and fears of pestilence, famine and wars.

Our consciousness is unfortunately a two-edged sword; it is forged both of hopes and fears and our efficiency in communication always dwells more on the latter, hence perhaps the tales of caution.

Chapter 9

Extinction is forever

One of the most important facets of evolutionary fitness is that the population of an organism is controlled in relation to the resources available to it. This is particularly well illustrated by the population of animals which live on small islands.

The chapter compares and contrasts population control in the relict population of Giant Tortoises on Aldabra with that of the human population on Mauritius, the most densely populated land mass in the world. It concludes that although survival may be for a limited time, extinction is forever; and that the small island of Mauritius has a lot to teach the world.

There are, even today, a few magical places left in the world where man has had little effect on the natural course of evolution. One of these is the island of Aldabra in the Indian Ocean where I was once privileged to work as a member of phase 6 of The (British) Royal Society Expedition. Islands have held a special fascination for biologists ever since Darwin's study of the Galápagos helped him to crystallize his ideas of evolution. You can, therefore, guess my excitement at being able to visit Aldabra, which although situated on the other side of the world, shares many similarities with Darwin's islands. There is little doubt in my mind that many of the ideas which form the basis of this book were re-shaped by my experiences on that and subsequent expeditions to other islands. It would, thus, be wrong for me to complete this work without at least some reference to those experiences. However, before I do it seems an appropriate point to recap on what the Galápagos experience did for Darwin and for the theory of evolution.

In short, it showed him the importance of isolation. There he found relicts of past phases of evolution, like the Marine (*Amblyrhynchus cristatus*) and Land (*Conolophus sub cristatus*) Iguanas and the Giant Tortoise (*Geochelone elephantopus*) living safe from competition and predation by mammals which had wiped them out on the adjacent mainland of South America. He realized that the ancestors of the Galápagos Cormorant (*Nannoptenum harrisi*) must have flown to the islands and, faced with no predators and a surfeit of food, forsaken the energy consuming power of flight for a new form and a new way of life. He recognized differences between and within the populations of closely related animals that lived in isolation on the various islands and, perhaps most important of all, he saw the finches. There were fourteen separate species of finch, each one adapted to a different aspect of island life by its stature and, especially, by the size and shape of its beak; fourteen permutations of tree-, cacti- and ground-dwellers and seed-, fruit- and insect-eaters, including one that uses a cactus spine as a tool to winkle out its food. Yet from their basic similarity he realized that they

Darwin's finches

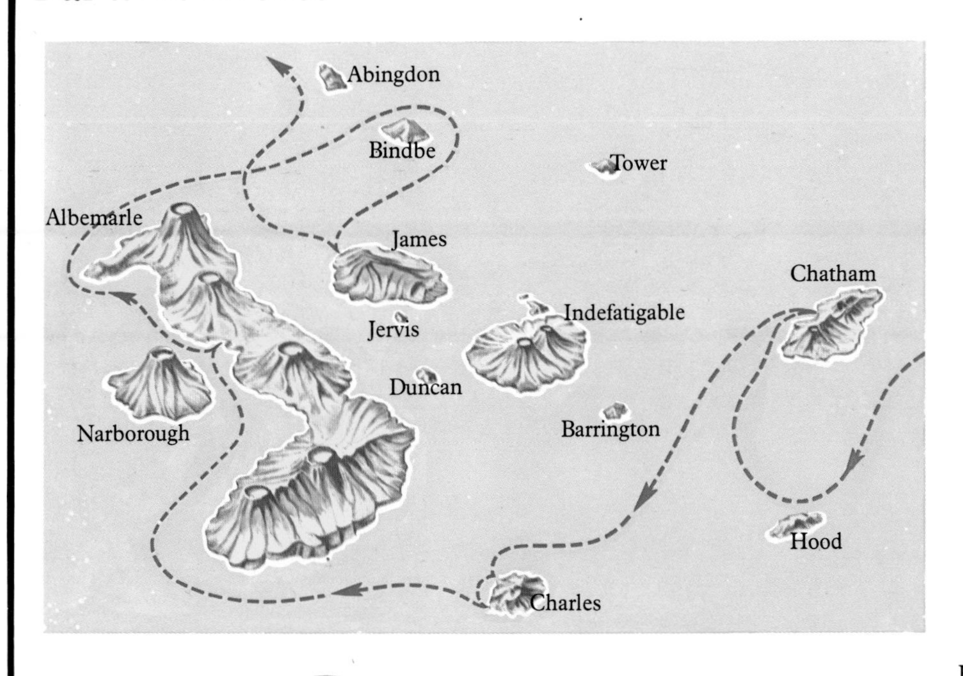

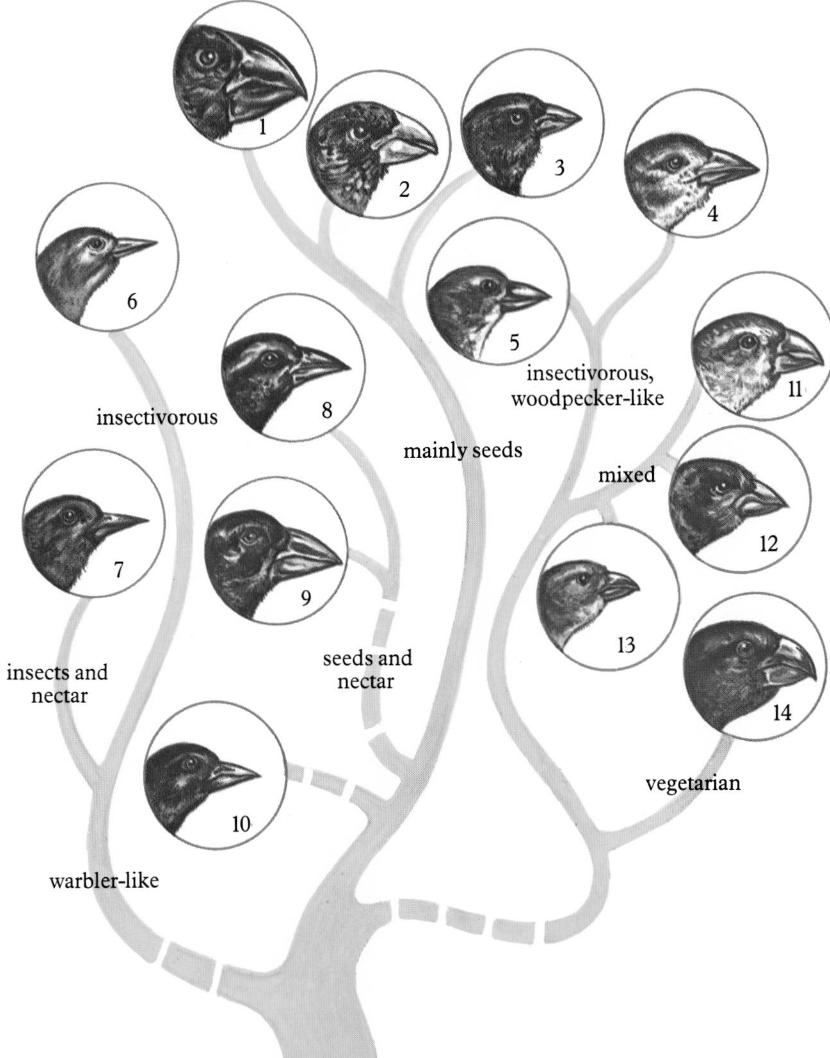

KEY
1 *Geospiza magnirostris* Large Ground-finch
2 *Geospiza fortis* Medium Ground-finch
3 *Geospiza fuliginosa* Small Ground-finch
4 *Cactospiza heliobates* Mangrove Finch
5 *Cactospiza pallida* Woodpecker Finch
6 *Certhidea olivacea* Warbler Finch
7 *Pinaroloxias inornata* Cocos Finch
8 *Geospiza scandens* Cactus Ground-finch
9 *Geospiza canirostris* Large Cactus Ground-finch
10 *Geospiza difficilis* Sharp-billed Ground-finch
11 *Camarhynchus psittaculus* Large Tree-finch
12 *Camarhynchus pauper* Charles Tree-finch
13 *Camarhynchus parvulus* Small Tree-finch
14 *Platyspiza crassirostris* Vegetarian Tree-finch

Darwin's observations, made as the *Beagle* cruised through the Galápagos, crystallized many of the ideas which were to make him famous. Over the preceding million years another warm-blooded animal had been 'cruising' through the island group and, finding no competitors, had got on famously, taking up all the food potential on offer.

Adaptive radiation, natural selection, specialization – it all seems so painfully obvious now it has been pointed out. I wonder how many of us would have had the ability to fit the facts and come up with a working model of evolution. My first glimpse of Darwin's finches was a medium-sized ground dweller who was busy feeding on the placenta of a newly born sea lion – wherever there is potential!

had all been derived from a single ancestral population that had probably been blown to the islands by some freak storm.

'Fourteen separate species'–in those words lay all the importance of isolation. Probably less than a million years ago, pioneer seed-eating finches arrived on the islands and, finding no competition, explosively took over the seed potential of the whole group. As the population increased, competition for the seeds gradually displaced those less well-adapted to make use of other sources of food. Natural selection working on the variation present in the genetic make-up of the population produced sub-populations which could make more efficient use of the various food sources on offer. Thus, a gradual change took place that opened up more and more of the food potential to the 'diverging' population of finches.

At first it was probably a slow process as interbreeding between the diverging lines would have been possible, diluting any change. However, as differing feeding territories, breeding times, behaviour and habitat began to isolate the sub-populations, interbreeding would be cut down, channelling the flow of genetic information and speeding up the process of change. It is possible that the main changes took place within sub-populations temporarily isolated on different islands or on different parts of the same island, the successful products then migrating to the new potential of other islands in the group. It is also possible that volcanic activity wiped out some of the evolving populations. Such chance happenings could have either speeded up or slowed down the process depending on which part of the evolving milieu was destroyed. The core of the whole process had little to do with chance. The potential was there and in time fourteen species evolved to make use of it; fourteen species isolated both in terms of their energy requirements and by the erection of barriers which gradually prevented interbreeding. Such isolating mechanisms may be geographical, anatomical, behavioural, genetical or biochemical, but once effective a new species has its origin.

'The Galápagos, a living labora-

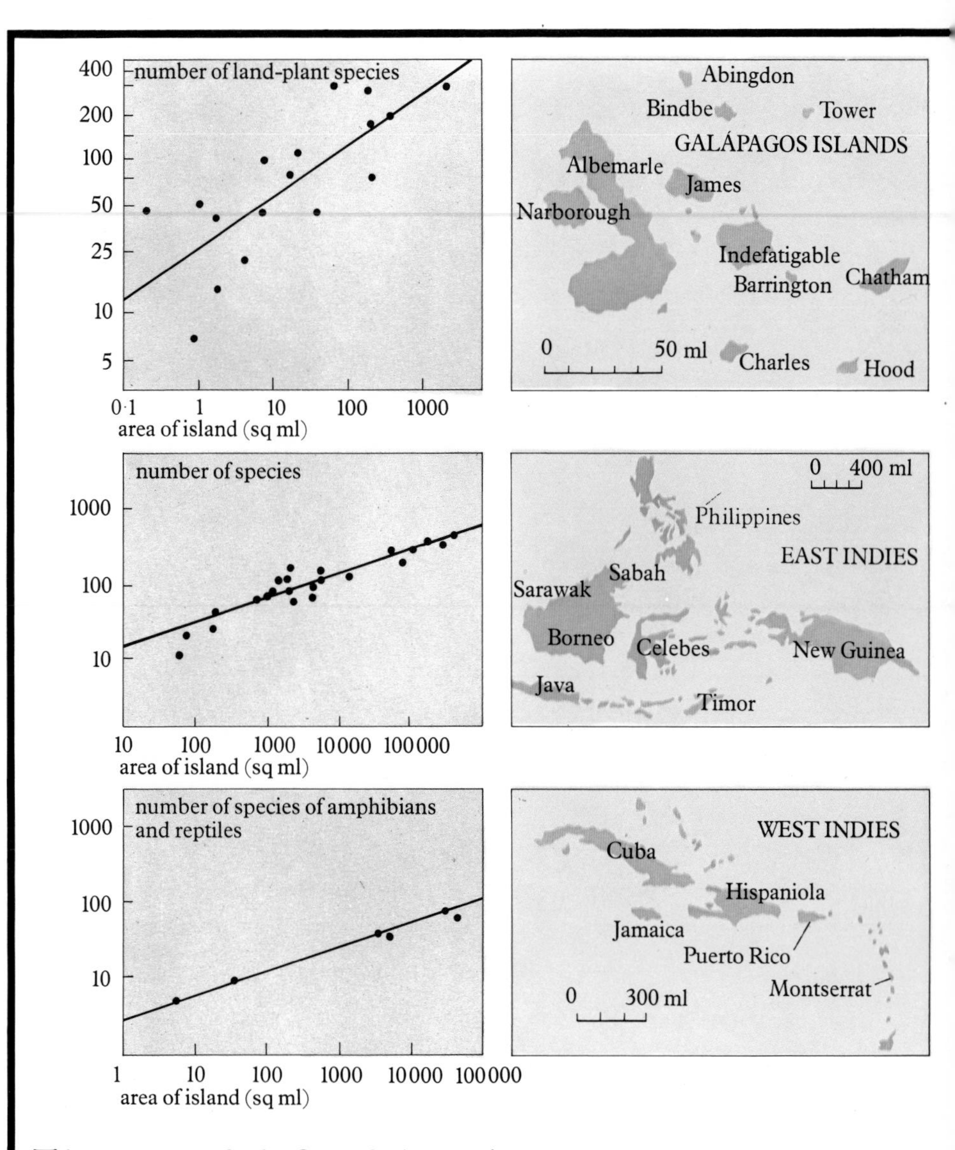

The potential of an island home

Logs may well have helped certain animals to make the journey from mainland areas to nearby islands just as they helped Thor Heyerdahl in his Kon-Tiki travels. Logs of a different kind have helped scientists to unravel some of the secrets of island life. Have they discovered a law of ecology? Only more research will tell, but if they have the law could be stated something like this: all other factors being equal, the potential of any land mass for colonization by plants and animals depends on its size.

tory of evolution'–is a phrase often used to explain how they clinched Darwin's ideas. In fact, they are no more a laboratory of evolution than any similar area of mainland. The difference is that on islands the picture is simplified by isolation.

Take a group of islands of varying size and plot the number of species present on each against the area of the island on a logarithmic scale. The surprising thing is that it does not seem to matter where your islands are located–tropics, temperate or even Antarctic waters–the resultant slope of the graph is the same. This does not mean that all islands the world over that are of the same size have the same number of species living on them. Tropical islands, provided they get sufficient rainfall, are always much richer in life than islands of the same size in higher and lower latitudes. Also, islands way out in the middle of an ocean will be poorer in species than those close to the mainland. The similarity is in the overall slope of the graphs and that means that in any biogeographical zone the number of species inhabiting an island will depend on the size (potential) of that island, which is limited, hence the simplification.

The species diversity of any finite area is made up of two opposing time-dependent processes, immigra-

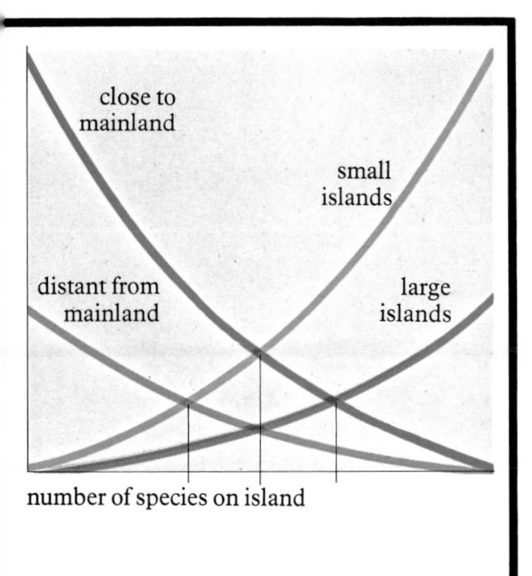

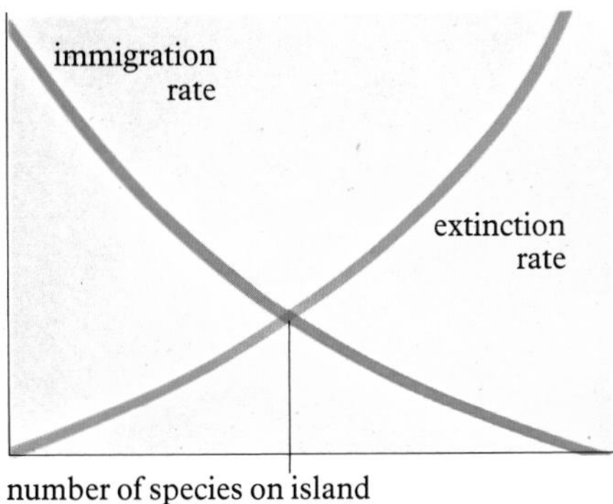

In recent years, the biological sciences have become more and more mathematical. The reason is two-fold: firstly, biologists have now collected sufficient basic data to allow them to ask complex questions and seek logical answers, and secondly, many mathematicians have found out how interesting biology can be. The most recent phase of biomathematical endeavour has been in the field of model making. The diagrams show the MacArthur-Wilson Equilibrium Model. They are self-explanatory and you might at first think that they are no more than a painful elaboration of the obvious. Think again, because once you have a model that works you can start to make predictions.

tion and extinction. Imagine a new island located at a certain distance from a large land mass, the flora and fauna of which comprise the pool of species from which the island will be colonized. The first species to settle on the island will be those best able to make the crossing and gain a foothold, and whose population and reproduction rates are greatest on the mainland. As more and more of the potential colonists are found on the island, fewer and fewer of the new arrivals will be new species; the rate of immigration, thus, gradually declines. As the process continues, more and more of the available space will be used up, patterns will stabilize, communities mature and the early pioneers will begin to be eradicated, and the rate of extinction will increase. Equilibrium will be reached when the rate of extinction balances the rate of immigration. At this point, the species diversity will become fixed, and it is determined by the potential of the island.

You do not have to live on an island to realize just how much potential and size are linked. The larger an island, in any dimension, the greater will be its 'capacity' for intercepting rainfall. The more water there is flowing down its ample flanks, the greater will be the chance of erosion producing a variety of landscapes, and weathering producing a variety of soils. Such variety is the spice of community life, and each community will be composed of many different species.

The only real problem is to know at what size an island ceases to be an island and, at least evolutionally, takes on the guise of a continent. Certainly Australia, with its relict populations of marsupial (pouch-bearing) mammals, must enjoy some of the benefits of isolation from the main stream of evolution. However, the success of her own special lines of advance, like the gum trees and kangaroos, show that with sufficient potential evolution can continue apace. Perhaps the speciation of the finches on the Galápagos show us that there is no need to draw such a distinction, and, after all, each continent is surrounded by water and the earth itself floats in space.

The equilibrium model, as set out above, may be applied with suitable modifications to predict the diversity of any land mass. The main modifications are that the larger the area under consideration, the more is the rate of immigration enhanced by the rate of production of new plants and animals (speciation), until for the whole world no such distinction can be drawn. The equilibrium model may appear to be no more than a painful elaboration of the obvious, but it is useful in that it forges two very important links. It links the two dogmas that only the fittest will survive and that wherever there is potential it will be used. It also links the two processes, the evolution of organisms and the evolution of ecosystems, showing that they are part of one and the same process and are thus subject to the same limitations.

One of the fascinations of Aldabra is that we know that at some time in the past the island did not exist and that it has never been part of a mainland mass. Aldabra is a coral atoll, which is not an island in its own right as it owes its existence to the life processes of a whole group of invertebrate animals, best called the reef-forming corals. These reef-formers enjoy a rather restricted environment, although it is one that many other organisms might envy. It is the upper part of the euphotic zone in sea water between the temperatures of 15 degrees Centigrade and 39·5 degrees Centigrade (59–103 degrees Fahrenheit). Perhaps part of the reason for the lower temperature limit relates to the fact that the same habitat in the colder waters of the world is occupied by large seaweeds (algae) most of which do not thrive above 15 degrees Centigrade and will, therefore, not compete with the coral. Their restriction in depth is also due to the presence of an alga, but in this case a minute unicellular alga that lives within the tissues of the coral. The coral provides the alga with a home and an enveloping supply of carbon dioxide, while in return the alga keeps the coral supplied with some of the oxygen it needs, at least while the sun is shining. In the absence of the alga, or insufficient light for its photosynthesis, the coral can not build its protective skeleton. This is a perfect example of symbiosis on a massive scale, for together these partners have built the reefs and atolls of the world.

Like the rocks of the Galápagos, the foundation of Aldabra is volcanic in origin, a platform of basalt thrown up as part of the seismic activity which is still forming the Indian Ocean. The basalt must have welled up into waters shallow enough to support the growth of the reef-formers which completed the job by producing the atoll. This platform raised up above the surface of the sea, some 1 000 kilometres (620 miles) off the coast of Africa, had its potential open to anything that could make the crossing. Today its flora consists of 269 species of vascular plant, many of which differ enough from their near-

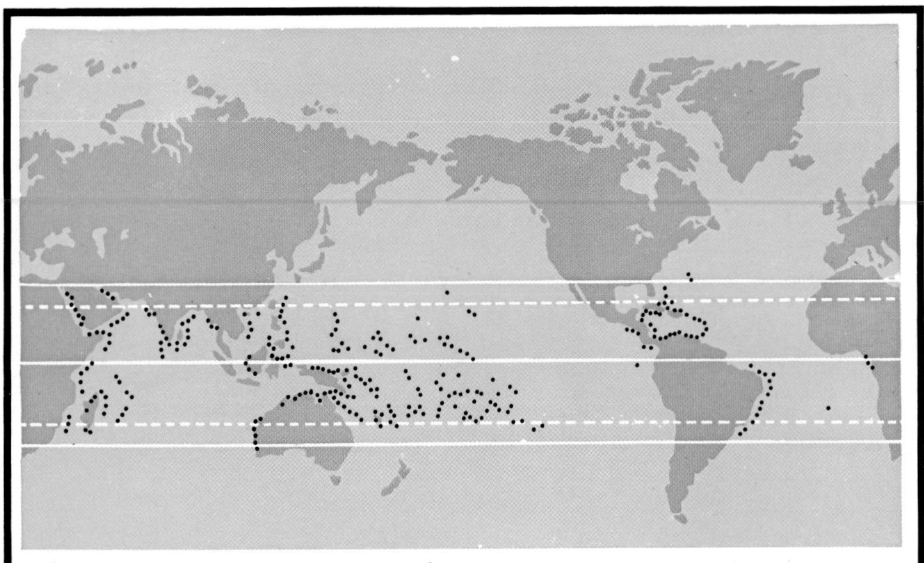

The reef-forming corals

Of all the animals in the world the ones with which I would be most willing to swap habitats are the reef-forming corals. The dots show the distribution of the reefs and reef formers of the world. On second thoughts, perhaps it is better to be a mammal because as such I can roam the habitats of the world.

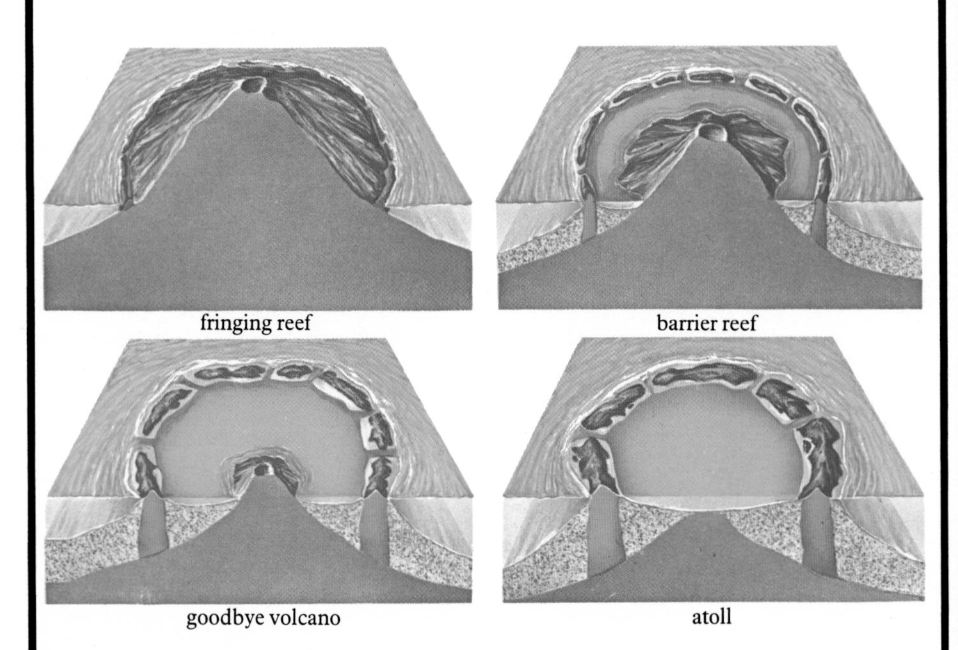

Darwin's theory of atoll formation

Charles Darwin must have also been fond of the habitat of the hermatypes (reef-formers) because the *Beagle* spent much time in their waters. His theory of atoll formation has stood up to the test of time and is finding corroboration in the new ideas on continental drift. A volcano protruding from the coral seas would make an ideal platform on which the corals could grow, forming a fringing reef. As the coral animals require lots of oxygen and do not like the water to be too hot, coral growth would be mainly limited to the ocean side of the reef. If the volcano gradually sinks into the sea (and now we know that the ones along the mid-oceanic ridge will do just that as the ocean floor spreads), the fringing reef will become a barrier reef separated from the shore by an ever widening stretch of water, and eventually form an atoll complete with a ready made lagoon. The whole thing could, of course, happen without the volcano ever appearing above water. All that is required is a platform of basalt, the surface of which reaches up into the euphotic zone.

Right Hermit crabs and the polyps of *Astreopora*, a reef-forming coral.

Overleaf The reef-forming coral provides protection for many thousands of different animals.

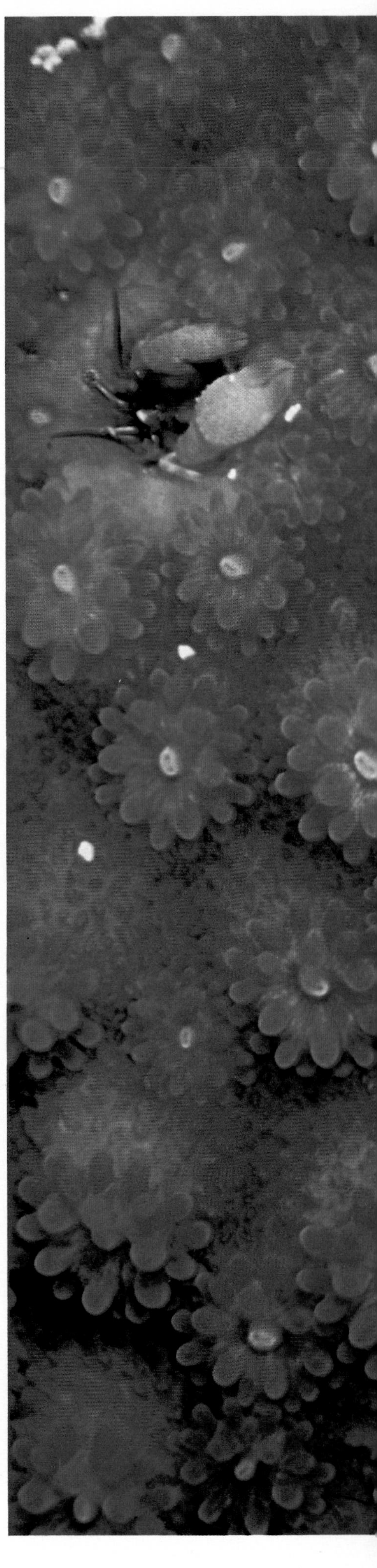

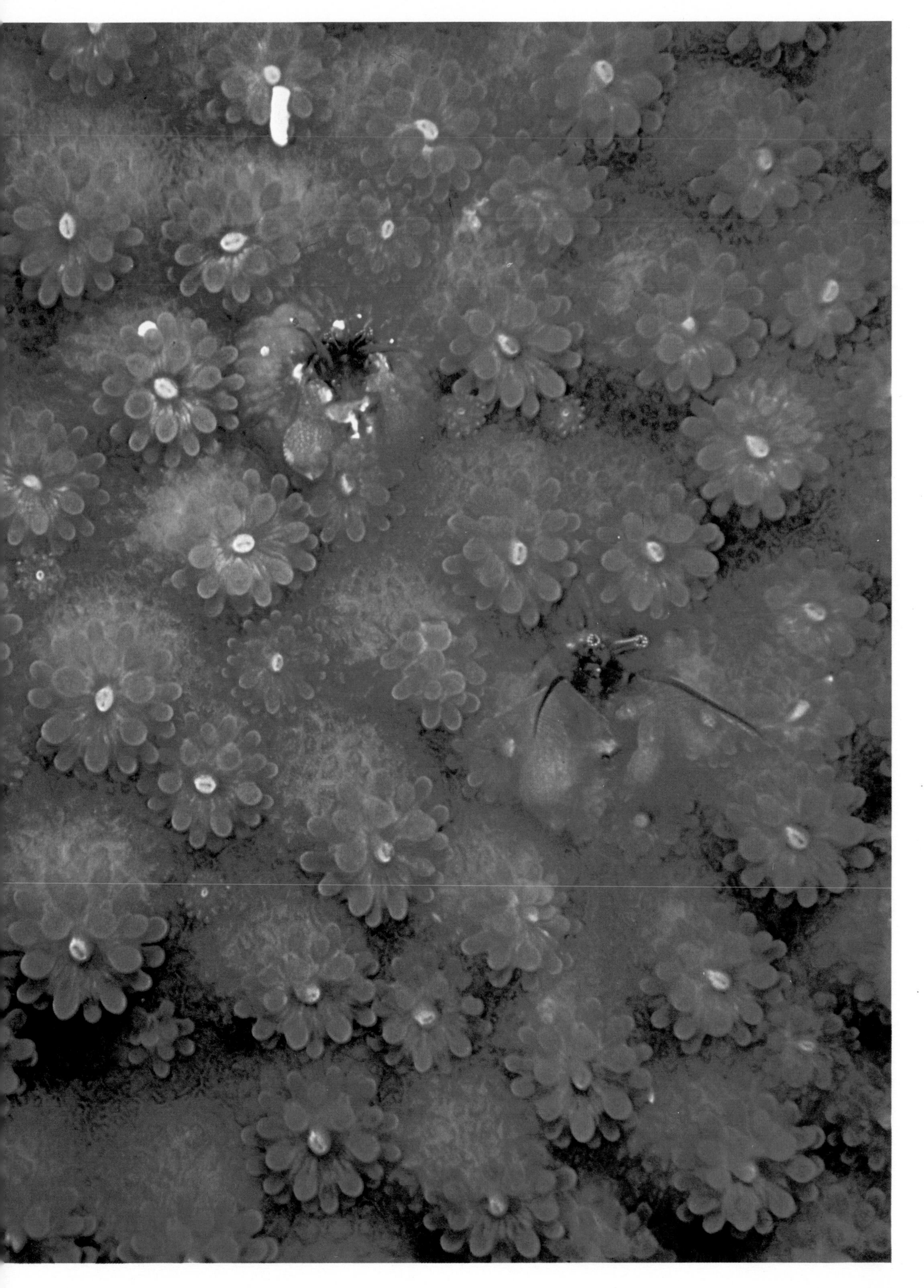

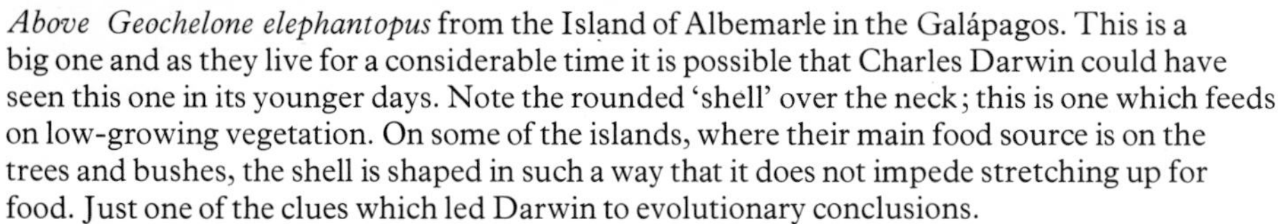

Above *Geochelone elephantopus* from the Island of Albemarle in the Galápagos. This is a big one and as they live for a considerable time it is possible that Charles Darwin could have seen this one in its younger days. Note the rounded 'shell' over the neck; this is one which feeds on low-growing vegetation. On some of the islands, where their main food source is on the trees and bushes, the shell is shaped in such a way that it does not impede stretching up for food. Just one of the clues which led Darwin to evolutionary conclusions.

est relatives on the adjacent land masses to be called endemic, showing that evolution is still going its own divergent way on Aldabra. Its fauna includes a flightless bird, the Aldabran Rail (*Dryolimnas aldabranus*), and a thriving population of Giant Tortoises (*Geochelone gigantia*).

One of the jobs I had been given while on the expedition was to repaint the numbers on the backs of a sample of Giant Tortoises which had been marked in order to gain an estimate of total numbers. The method is a modification of the Humphry Davy dilution principle, originally used to determine the volume of coal mines. In the case of mines, a known volume of an inert gas was released into the mine. After thorough mixing, the concentration of the gas in the mine atmosphere was measured, the dilution giving a measure of the total volume of the mine. In the same way, the group of marked animals released into the population provides an estimate of total numbers.

It was great fun locating the animals and then sitting side saddle on their backs, paint brush in hand, endeavouring to repaint the numbers upside down as you rode across the rough terrain. It was this close contact with these relicts of the past which raised a number of questions in my mind.

First, in Darwinian terms here was a successful animal, for it had survived almost unchanged for millions of years. I was very tempted to paint a capital D on its back–passed by Darwin as fit to survive. However, it must be remembered that during this time its population had become smaller and smaller and its territories more and more restricted as it was ousted by later products of evolution. Is isolation the only secret of its success?

Analysis of the resources of the atoll in Tortoise terms indicated that none of them were in superabundance. The irregular terrain, with many pits and holes, was not the ideal habitat even for such a slow mover. Rain falls on the island only at irregular intervals and when it does, in most places, it quickly drains away through the porous limestone. Where pools of fresh water do exist they are centres of Tortoise activity, but over much of the island the reptiles must derive much of their water from their food.

The vegetation of the atoll consists of a variety of scrub and scrub forest, interspersed with more open areas, especially around the sand dunes, the latter being much favoured by the Tortoises as a source of easy grazing. In places, signs of grazing pressure ranged from close-cropped turf to Tortoises balancing precariously on their back feet while attempting to reach the topiarian trees. The scrub forest, however, not only supplies the Tortoises with food and water, but in places it forms their most important resource, shade. The importance of the latter is evidenced by the number of dead and dying Tortoises which appear to have been caught out in the open far from shade. It was a pitiful sight to see the great creatures with their heads pushed into some small shady corner but their vulnerable dark brown backs out in the full glare of the sun. Dark bodies, be they inanimate or alive, absorb heat with great efficiency, so much so that too long an exposure to the full glare of the midday sun means slow death to a Tortoise,

Above left *Geochelone gigantia*, living on the other side of the world from the Galápagos, isolated on the coral atoll of Aldabra. Darwin never visited Aldabra but the life on this small crop of land tells the same story of evolution, where only the fit survive. Part of the fitness of the Aldabran population lies in the fact that some of the members of the population stick to the shade, while others stick their necks out into the midday sun.
Top Too many Tortoises destroy their most important resource. On some parts of Aldabra, where the population of the Tortoises gets too dense, their efficient grazing effectively stops the regeneration of the same forest which provided them with shade throughout the hottest part of the day. Grasses, sedges and other grazing tolerant plants grow up in the clearings, tempting adventurous Tortoises to a fate worse than death.
Above One of the whiz-kid Tortoises that stayed out too late and paid the ultimate price. The heat of the tropical sun on the dark carapace of the Tortoise are a lethal mixture and if the Tortoise is caught out around midday it will be killed, baked in its own jacket.

causing it to be baked in its own jacket.

The whole thing appears much more ironic when it is remembered that the dark carapace (shell) of the tortoise evolved as a protection against predators. Could it be that the adaptation which protects the individual in one situation plays exactly the opposite role on its island refuge, and yet in the long run protects the species from extinction? If the Giant Tortoise population on one sector of the island became too large (too successful?) and started to destroy its most precious resource by preventing regeneration of the scrub forest, more and more of the population would die of exposure. The population would thus be cut back, regeneration of the scrub forest could start and the Tortoise would be saved. Just how important this has been in the long-term history of the Aldabran Giant Tortoise I do not know, but I am sure that without some form of natural population regulation the Tortoises would be no more.

One other fascinating fact is that I think I have sat and watched a very special form of natural selection in operation. If what I have reported above is true then, the unadventurous Tortoise, which never moves far from shade, even if it means living on a meagre diet, is much more likely to survive than the adventurous whiz-kid Tortoise, always roaring off in search of pastures new. Perhaps there is no such variation in the population but if there is, or if there has been, here is a case of the meek inheriting the earth while the aggressive go to the wall as part of the key process of survival.

The success story of the Giant Tortoise is written all over the atoll in the form of bones bleached white in the sun. Suppose a breakthrough was made and a race of super-tortoises with white shells that reflected the heat of the sun began to take over. Would they destroy the forest and themselves, or would some other limiting factor come into play? Again I do not know the answer. What I do know is that although survival may only be for a limited time, extinction is forever.

Of all the animals that have been made extinct by man, the most famous is *Raphus cucullatus* for its demise is recorded as a catch phrase of the English language, 'as dead as a Dodo'. The Dodo's native home was Mauritius. Comprising 1865 square kilometres (720 square miles) of volcanic rock, it is the second largest of the Mascarenes. Of all the islands in that area, the flora and fauna of Aldabra have suffered least at the hands of man. The reason is not too hard to understand. Its climate is too dry for commercial copra and it has no exploitable deposits of guano, for its large populations of sea birds nest in the mangroves and void their desirable waste into the tidal waters of the lagoon. In short, it is an island that offered little or nothing in the way of potential to the late immigrant man.

'As dead as a Dodo'. This one is unfortunately only a reconstruction. Once common and widespread on the Island of Mauritius, this large flightless bird was unable to cope with the coming of man. Hunting, destruction of habitat, and most important of all the introduction of predators and competitors, made the Dodo extinct. As you land at Plaisance Airport for a fabulous package holiday, just think that the Dodo once roamed the bush on which the airport is built.

The same is, unfortunately, not true of the majority of the islands in the area and especially the large ones like Mauritius. When man first arrived on Mauritius on the 18th September 1598, *Homo sapiens*, in the guise of Admiral Wybrandt van Warwyck, a Dutch navigator, found an island covered with tropical vegetation. In the lowlands and fertile uplands, dense evergreen forests grew; at the highest elevations and on the poorest soils, a more stunted heath and scrub forest existed; while on the dry coastal flanks of the island, palm savannahs were found. All the early visitors commented upon the abundant and tame avifauna, which included among its numbers the large flightless Dodo and the smaller flightless Rail, now confined to Aldabra.

Once man colonized the island, exploitation of the valuable hardwoods, including ebony, was rapid and catastrophic, as was the slaughter of the edible birds for food and sport. By 1638 the Dodo was very rare and it was extinct by the end of the 1680s. The bulk of the lowland forests had gone by 1750, when extensive areas of land had already been put to agriculture. Such was the potential of the island to man.

Date	Acreage of native forest on Mauritius
1753	406 137
1770	388 705
1804	313 000
1836	300 000
1846	142 000
1852	70 000
1880	16 000
1936	7 000
1974	4 500

The destruction continued apace, and with the disappearance of the forest many endemic plants and animals disappeared off the face of evolution. It was, of course, not just the destruction of the forest which affected the native flora and fauna, but also the introduction of alien plants and animals by man. Man, the great destroyer of geographical barriers, introduced along with himself other non-flying mammals–rats, cats, dogs, mongooses, deer, monkeys and game birds, to name a few–all of which added to the picture of destruction. Similarly guava, casuarina, blackberry, privet and many others ran amok through the pitiful scraps of semi-natural vegetation which exists between the sprawling order of the sugar and tea plantations.

Today, it is true to say that no natural vegetation exists, although in the remotest and most rugged areas many unique plants and animals have taken their last stand against extinction. Perhaps the plight of the wildlife of Mauritius is best summarized by reference to the Dodonut tree and the remaining endemic landbirds. *Calvaria boutoniania* is a tree of moderate stature which produces a fruit of fantastic hardness. I have one in front of me as I write and the might of a modern ratchet nutcracker is powerless to crack its shell. Not so the beak of the Dodo, for legend has it that its giant bill alone could crack the nut, allowing it to germinate. To corroborate the story the tree is today very rare and is on the danger list. Research has shown that in the absence of the Dodo domestic turkeys make a good substitute, but in their case it is the digestive enzymes of the crop that soften the shell making germination possible, so perhaps the tree will not become extinct.

Out of the eight land birds which deserve the distinction of the title endemic, all are on the world list of endangered species and the populations of all are known to be declining. One of the eight, the Mauritian Kestrel (*Falco punctatus*), of which only nine specimens probably remain, vies for the dubious honour of the title of 'rarest bird in the world'. My favourite is the Pink Pigeon whose ample proportions, promise of pigeon pie and tameness, must have put it high on the list of first to become extinct. Its saving grace, if such it could be called, was the fact that the bird is often 'high' on a narcotic produced by one of its favourite food plants *Euphorbia pyrifolia*, a habit that makes it poisonous to man. Today there are only a few specimens left, due mainly to habitat destruction and predation by monkeys, which appear to thrive on a second-hand 'trip'.

So it would seem that the locals are not doing at all well, but what of the colonists? Discovered in 1598, by 1959 there were more than 600 000 humans (that is, more than 300 to every square kilometre) jammed on to the island. This is success by any standard, and the growth of 3 per cent per annum was almost exponential.

The secret of this success was an ebullient world trade demanding sugar and tea, the two products which are specially favoured by the local climate. The flow of cash into the island lifted an agrarian society up to enjoy at least some of the benefits of the industrial medical revolution. Mechanization has made both the sugar and the tea industries less dependent on the sweat of human labour. The import of ideas, expertise, education and drugs has controlled the rate of death and suffering, and has, especially since the 1940s, reduced the rate of infant mortality. The efficacy of the latter is shown in the make-up of the island's population, 44 per cent of which were under the age of 15 in 1959. This imbalance could not be a bad thing in terms of the future population, provided that the resources were available to support such growth. Unfortunately, that depended on the continuing growth of world trade. The fact that it was not happening was written all over the island: slum conditions were rife; impossibly overcrowded primary schools in some places worked double shift systems; six out of every seven students from most secondary schools, who could not find a place in high school, swelled the number of unemployed, many of whom sat waiting for the promise of the next sugar harvest. Political and social unrest were a feature of the news.

The government responded boldly with new and imaginative schemes– the harnessing of wave power on a massive scale to provide power which is not based on imported fossil fuel; a large wave-free area in which fish farming will boost the island's supply of edible protein; new forestry developments which aim to replace some of the areas of scrub by productive plantations of Slash Pine; the encouragement of tourism by the enlargement of the airport, the construction of hotel complexes, and the opening up of the best areas of beach, reef and landscape for the enjoyment of hard-cash visitors. Unfortunately, all this must put greater pressure on the dwindling areas of semi-natural vegetation and the wildlife.

Perhaps the new breed of tourist who wants to see the natural beauties and treasures of the land they visit will ensure that the bounds of the new national parks and reserves are not violated. But these are long-term benefits, and what would you do if you were a local faced with the immediate problem of feeding your family? Would you not, perhaps, decide that a Pink Pigeon in the pot is worth more than gratuitous tips in the future, especially when they are based on an as yet unproven increase in tourist potential.

It is much easier to take the money to be made planting pine trees or pioneering a tidal barrage than to heed the words which caution that both may be swept away by the next big cyclone. Survival is the name of the game and, as extinction is forever, with nothing better on offer the only thing to do is to grab that which is immediately available and hope, like the dead Tortoises, that the corner of shade is large enough to contain all your requirements.

It would, perhaps, be worth waiting if you knew that help was at hand, or some undreamed of breakthrough was about to take place opening up more potential. If only the problem of too little potential for too many humans was unique to Mauritius then help would be more readily at hand. Unfortunately, the symptoms outlined above are those of a worldwide syndrome stemming from the unprecedented success of man in recent years. In the knowledge of this the Mauritian Government officially recognized the need for population control in 1965, just one of eleven countries so to do in the same year. The winds of the next breakthrough were beginning to blow.

Below The rarest bird in the world, *Falco punctatus*, in protective custody. But stop press news – it is starting to nest on steep sea cliffs out of harm's way, so it may be making a comeback all on its own.
Overleaf The south-west of Mauritius, showing typical weathered volcanic peaks in the distance. The unprotected south coast of Mauritius is perpetually buffetted by waves. If this energy could be harnessed to produce electrical power, the islanders would not need to use expensive fossil fuels.

Chapter 10

Spaceship earth

Wherever there is potential it must be used – twentieth-century man is obeying pre-eminently this rule, which has 'directed' evolution from its physico-chemical beginnings. We use more geochemicals and more water and degrade more energy and more environments than any other species. The potential was there; the rules had to be obeyed; evolution had to happen from photosynthesis through the 'explosion' in the Cambrian to the industrial/medical revolution. Each stage has acted as a positive feedback speeding the process on. The only development that appears to be out of line with the rest of evolution is the new consciousness of concern. Is this the meaning of man's free will? And has it come in time to gain for him the final stamp of evolutionary approval, survival?

I believe it has, but I am an optimist.

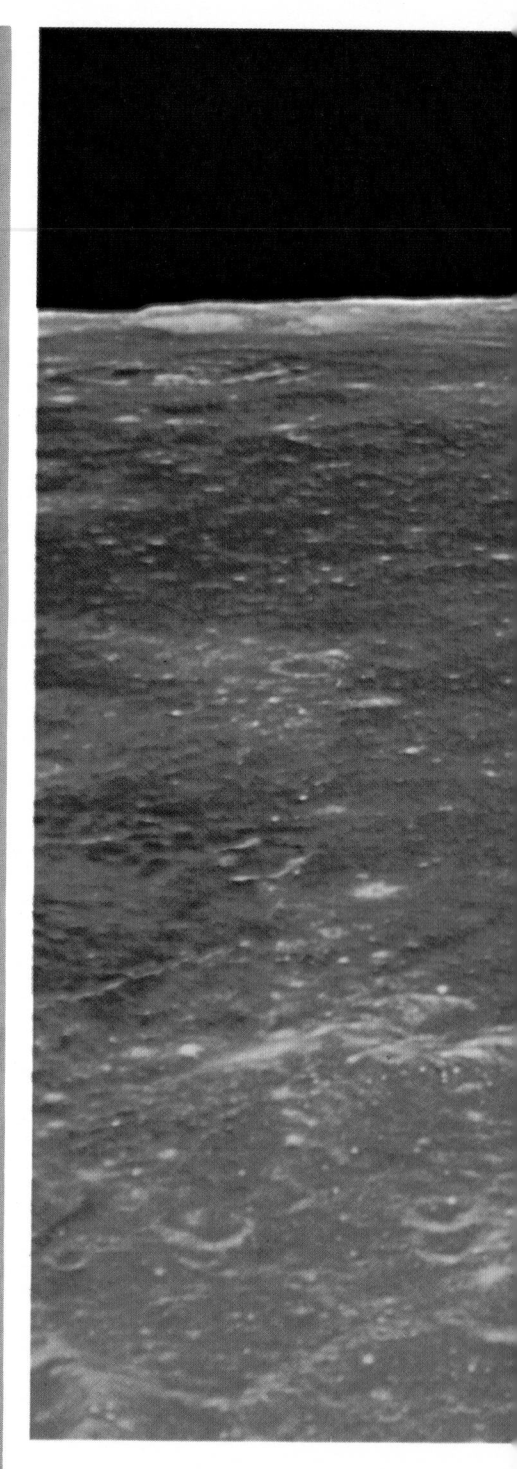

On 21 July 1969, Neil Armstrong became the first man to set foot on another part of the solar system. In his own words 'it was a small step for a man, but a giant leap for mankind'. He was the first of the new explorers, seeking potential beyond the confines of the earth, and through the 'eyes and ears' of the most sophisticated communications network then known he shared the experience with countless millions of earthbound humans.

The Apollo 11 mission did not find that the moon was made of best blue cheese or some equally exploitable resource, with the promise of a new phase of colonial affluence. What it

Two space capsules seen out of perspective, a view that for millions put their world in perspective. The Apollo 11 lunar module over the moon with your world in the background. All systems go.

did do was to prove that technological man is now fit for space and anything it has to offer, and, perhaps most important of all, it put the earth in its right perspective, a tiny speck floating in a vast sea of space

Everyone who followed the mission in detail knew that this was the most expensive step which man had ever made, and yet it all depended on the strictly limited resources of the space capsule and its life-support system and on the knowledge and skill of the crew in handling it. The earth looked small from the capsule, yet in the same way it was the life-support system for 3500 million people, all of whom are equally dependent on the knowledge and the skills of the people who handle it.

The society that put the first man on the moon could be said to be the most successful, a fact that is certainly supported by the magic statistic of per capita gross national product (GNP for short). This is the total sum of personal and governmental expenditure on goods, services and investment divided by the population. In essence, it is the actual amount of money spent by a country (the cash flow) worked out on the basis of every member of the population having an equal share. In 1969, the year of the Apollo 11 countdown, the United States led the world with a GNP of 3980 US dollars. The total cost of the mission, spread out over 5 years, represents an investment per head of only 20 US dollars per year, which is about the amount that the average American citizen spent on hair lotions, lipsticks and other cosmetics. No wonder they could afford it, and it was well worth the invest-

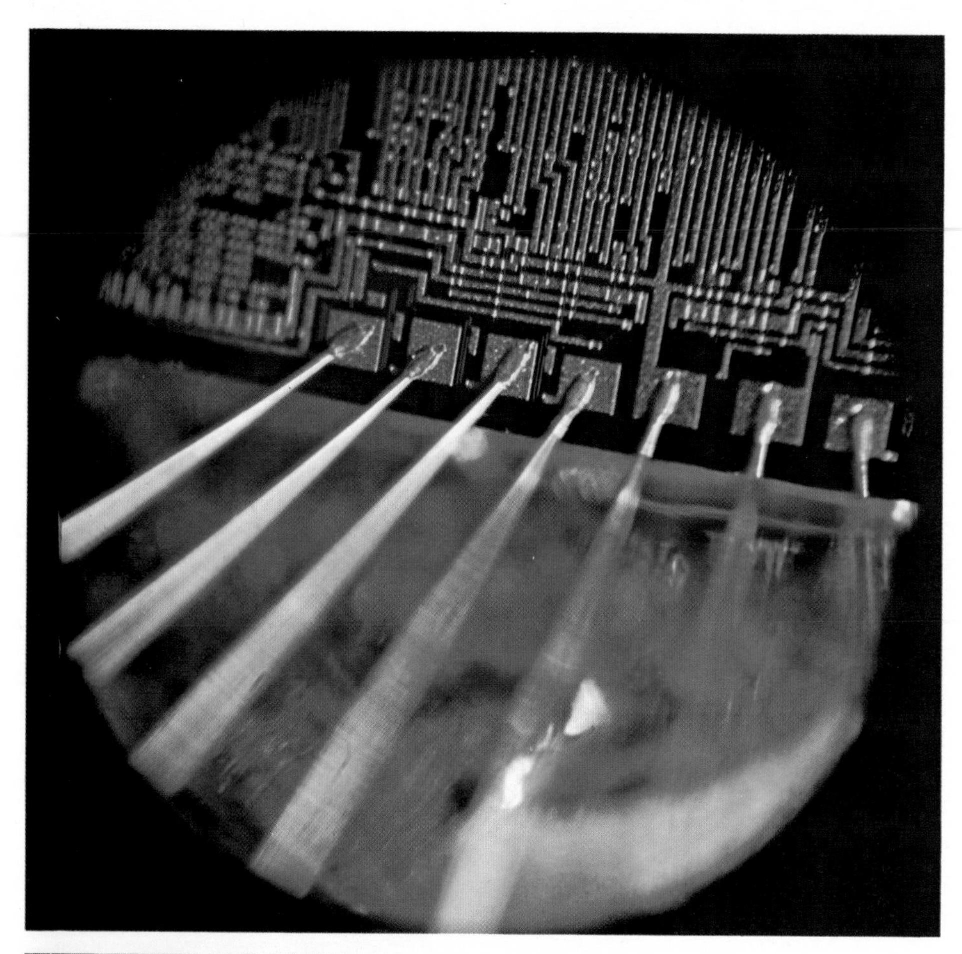

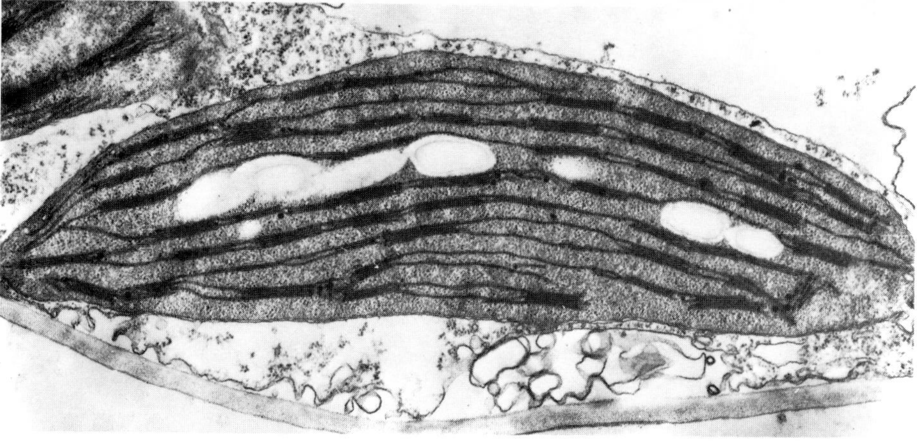

Top A close-up of the corner of a silicon chip on which modern electronic engineers have built a microcircuit (× 150).
Above The chloroplast of a green plant as seen under the electron microscope (× 17000). Piles of chlorophyll rich grana lie in a granular stroma and together they drive the living world.
The chloroplast is an energy converter and like the cells of our bodies it is made up of biochemicals with a backbone of covalent carbon. The microcircuit was designed by the 'brain' of man and engineered with his dextrous hands. It is made of silicon, another element which like carbon has the ability to form covalent backbones. Could this be the first step in a take-over bid, or will the new breed of energy converters and information sorters now being built into the micro-computers help us to build the models that will show us how to save the living world?

ment, not only in prestige, educational and entertainment value, but also in the multitude of technological spin-offs that have helped to make life easier and have solved a number of major problems.

Consider, for example, the wireless set of the good old days. It certainly never lived up to its name, for it was overflowing with copper wire of all gauges. If the systems of Apollo had been built on the same pattern, the weight of copper alone would have held the mission firmly on the ground. Miniaturization has been the key to the space age, and miniaturization has saved enormous amounts of precious raw materials as well as giving us 'invisible' hearing aids, pacemakers and so on, and the mixed blessing of transistor radios.

Necessity has always been the mother of invention, and so the foundations of the space race were laid in World War II, when technological advance became the key to survival. During those fateful years the best brains of the world were focused on solving the problems, and the best technicians turned their ideas into working reality. The war accelerated advances on a massive scale in a number of important areas – medicine, communication, cybernetics (control systems), transport, weather forecasting and survey, to name but a few – and out of the holocausts of Hiroshima and Nagasaki was born the hope of a new source of power.

The millions who died in the war were soon replaced as the new medicine, based on antibiotics, synthetic drugs, mass inoculation and a realization of the importance of hygiene and medical care, slashed the rate of infant mortality and pushed up the expectations of a ripe old age. At that time the promise of atomic power seemed, perhaps, most significant because the war machine had put an enormous drain on the reserves of fossil fuel. It is so easy to forget that the whole of the industrial, medical and technological revolutions have been fuelled by the excess of past ages of photosynthetic affluence. All flesh is grass, and all man's advances have been fuelled by plant products (well, at least 99·99 per cent of them have) and that is surely enough to earn us the title of 'botanic man'. Was atomic power going to be the shape of things to come? Was this the end of the era of power from plants?

Of all the nations of the world, the United States and Soviet Russia emerged from the war with sufficient resources and motivation to spearhead the advances into space. The resources relate in part to the size of the countries, both of which straddle the main wheat-growing climates of the world and are rich in coal and/or petroleum reserves, and minerals such as ferrous iron and other metals. The motivation was there because the potential was in the able hands of two opposing brands of consciousness, capitalism and communism. As far as the space race was concerned they both did very well, one with a capital

GNP of 3980 US dollars and the other with a communal GNP of 1110 US dollars in 1969.

America may have beaten Russia to the moon, but in GNP terms she was in 1969 run a close second by the tiny oil sheikdom of Kuwait whose 800000 people should have each got 3540 US dollars in the GNP share out. This indicates the importance of oil in world economics and shows that the potential of a nation can not be judged by its size alone. It is, in fact, very difficult to demonstrate a correlation between GNP and any readily measurable attribute of a nation. All one can do is to say lamely that GNP appears to be related to the importance of the role that particular country plays in the global economy, either as a supplier of raw materials (what it has that others need), or a producer of manufactured goods (what it produces that others want).

For anyone living in one of the more affluent countries of the world, a trip around the local supermarket will be proof enough of the worldwide system of trade on which we depend and of which we are an integral part. Cereals from Canada, butter from New Zealand, sugar from Mauritius, tea from Assam, bananas from Ecuador–it is like a world cruise in a wire basket. I like to keep a map of the world and mark on it the source of all my food. By the end of the week I begin to feel like a very important person–a consumer sitting in the middle of a resource web that covers the world.

What do I do in exchange for all the food I eat and goods I require? I am a communicator, both at a university and through the media of television, radio, books, magazines and newspapers. You may say, 'what a useless job, the world could do without communicators'. You are, of course, very welcome to your own opinion, but be very careful if you work in advertising, electronics, printing, paper making, bookbinding, retailing, and so on, because your livelihood depends, in part, on the success of communicators. We all live as part of a complex worldwide system, the affluence of which now depends on the potential of the whole.

So it has been throughout evolution. In the simplest terms, there could not be animals until there were plants on which they could feed, or carnivores before there were animals to eat, or decomposers before there was something to decompose. The potential for one depends on the potential for the others, each being part of the total living system.

The first organisms were, thus generalists each able to perform all the tasks that there were to perform. As evolution got underway, more specialized organisms came into being, reaping the potential on offer and creating more for others. As the process continued a greater diversity of plants and animals came into being, those which could not stand the pace of change and could not get a share of the available resources becoming extinct, while the others went on to add to the expanding diversity taking up their places in the ecosystems of the world. The ceiling level for diversity is thus dependent on the fitness of the environment for life, measured in terms of availability of energy, water and life chemicals, and the fitness of the evolving milieu to make use of it.

The argument can be elegantly applied to any or all of the phases of the social evolution of man. As a hunter-gatherer there were a particular number of jobs to be done, skills to be learned, roles to be played, and each member of the

Botanic man

'All flesh is grass' and this illustration shows botanic man composed of some of his staple crops.

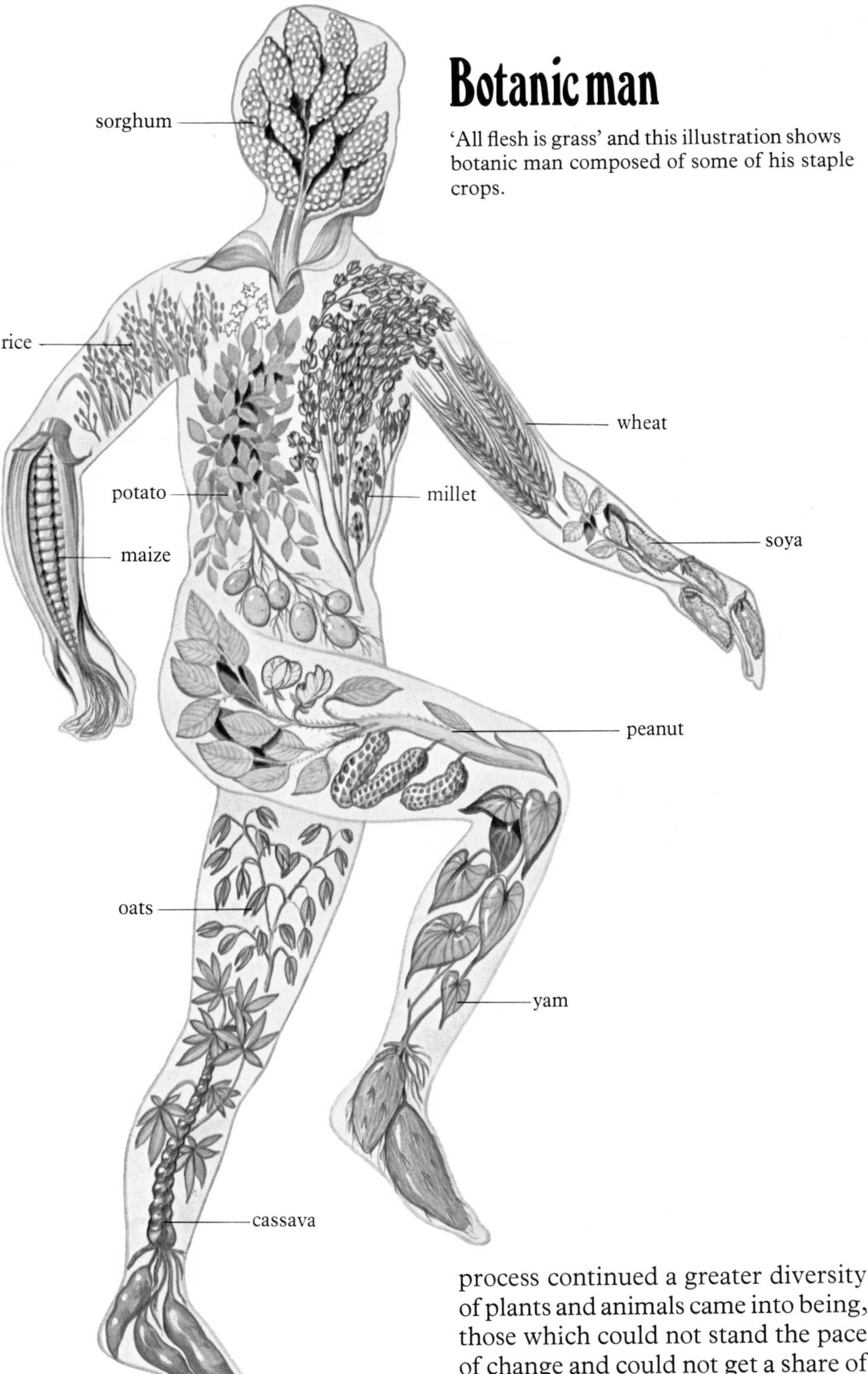

A medicare centre in Barakpore, India. One of the benefits which has come with the more modern way of life is a whole complex of medical centres throughout the world. The importance of such centres to the immediate welfare of the people cannot be overlooked. However, with efficient 'death control', especially the controlling of diseases such as malaria, the need for 'birth control' has become more and more important. India has led the world in attempting to organize family planning on a massive scale.

society was a generalist, learning all the necessary skills to live. That does not mean that some were not better at certain things than others, but that all of them had the ability to survive on their own. With the development of agriculture there was more potential, but only through the performance of a greater number of jobs. Specialists began to evolve and individuals became more dependent on the society and vice versa. The industrial revolution increased the potential and the problems, producing a more complex society in which more and more specialists became more and more dependent on the society and on each other. Again, we can see that the ceiling level for diversity is dependent on the fitness of the environment for the industrial life, measured in terms of energy including food, water and raw materials, and the fitness of the evolving society to make use of it. We are back to 'what you have that we need' and 'what you produce that we want', the laws of supply and demand that make the GNP go round.

Since the war the world population has been gradually rising and many people, biologists, demographers, economists and humanitarians, warned that shortages of food would soon limit the world population. These warnings were not new; in 1798 Thomas Malthus had warned in an excellent essay that 'man's power of population is infinitely greater than the power of the earth to produce subsistence'. He was the son of a wealthy country gentleman and had a good life, obtaining a first-class honours degree in mathematics from Cambridge, England, in 1788. He looked at the world through the eyes of a mathematician and came to the conclusion that population, when unchecked, increases in a geometric progression and that subsistence only increases in an arithmetic progression. From his position of personal comfort and in an atmosphere of great human optimism (for his was the 'age of reason', a time when learned and wise men saw themselves on the threshold of a world of concord among men and nations, in which want and oppression would not exist), he realized that the laws of nature condemned the bulk of mankind to live in the margin between barely enough and too little.

At about the same time the peasants of France came to the same realization, although theirs stemmed not from affluence and learning but from the stark reality of neither cake nor bread to eat. They effectively argued their point through the French Revolution, with the aid of Madame Guillotine.

This was, however, a time of great advance; the industrial/medical revolution was underway. There was more opportunity for employment, the promise of better living conditions, new potential in new industrial settings. All this was enough to lay the ghost of Malthus in at least a temporary grave, yet it was the medical advances that were to hasten the day of Malthusian reckoning. As the death control measures of the late 1940s started to bite the world population began to increase dramatically. World trade routes were once again open and agriculture responded to the challenge with more and better chemical fertilizers, the use of synthetic pesticides and herbicides, extensive irrigation, enlargement of fields to take the new heavy 'tanks' of agriculture, intensive production of chicken, beef, pigs, and so on. Similarly, the fishing fleets moved further from home to new waters with the promise of areas of upwelling. Fish and mammal populations which had for 5 years been under the protection of U-boat patrols were now hunted with the methods and devices used in submarine warfare. A boffin load of swords were fashioned into futuristic ploughshares.

These advances again mitigated against the scepticism of the new Malthusians for, on a world scale, food production kept pace with population growth. In fact, in the 1950s the international agricultural trade was complaining of burdensome excess, and American farmers were subsidized to take cereal acreage out of production. Yet in that same period food production in thirty-four countries failed to keep pace with population growth. Famine was only staved off in those countries by imports from North America. The grain drain was well underway.

In 1963, the imaginative World Food Programme of the Food and Agricultural Organization of the United Nations got off the ground. Its aim was not just a free handout of food. Food was supplied with one of the following purposes–to help a work force on a project that would aid their own agricultural prosperity; to provide for a sudden influx of

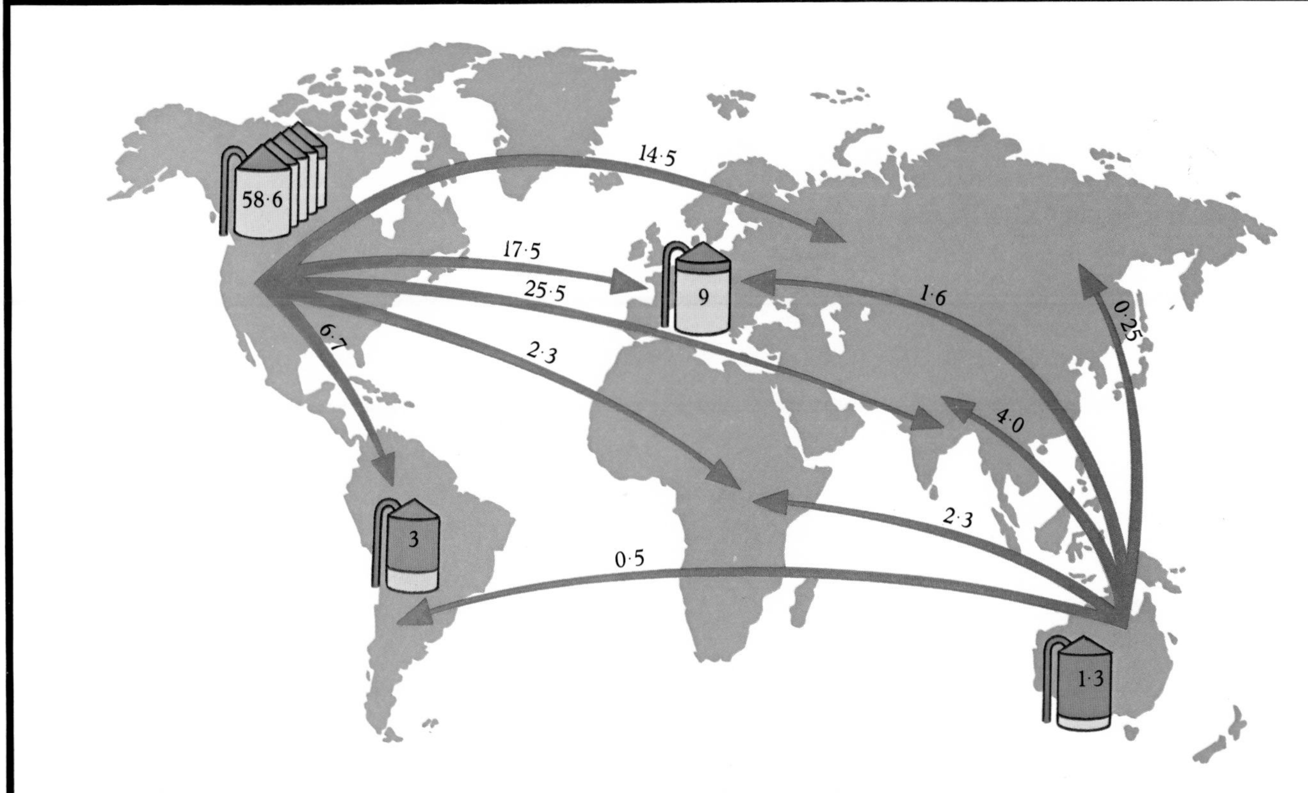

The brink of disaster

The net world trade and stock of cereals remaining in 1973 (figures in millions of tonnes).

Left Drought and the expanding deserts of the world. During the Sahel famine destitute nomads from Mali came to Christine Wells in northern Upper Volta in search of pasture but found a wasteland. Here a woman pounds in a mortar the last few grains in her family's possession.

Overleaf A crop of soya beans – hope for the future. The soya bean is a crop which cuts out the middle men since its bean contains many of the essential amino acids which were once derived only from meat. Removing grazing animals from the food chain puts us one step nearer the sun, which means more food for less land and hence more food for everyone. In addition, soya bean has nitrogen fixing root nodules and so cuts down the cost of fertilizer. Known as the meat bean in the east, it is now helping to feed more and more of the protein-hungry world.

refugees; and to safeguard against threatened famine or to succour the survivors when it did occur. There were, of course, enormous problems in putting the plan into action, not the least being to get the food to the right place in time for it to be effective, and to sponsor the right development schemes.

The consoling facts were that nations were beginning to work together in a common cause, and, on a world scale, there was more than enough food to go around. At least, the latter was true until 1973. In 1972, for the first time since World War II, the global production of food declined. Disastrous harvests forced both Russia and China to buy 30 million tonnes of cereal from North America. By the end of the growing season, the world stocks were so depleted that there were only 37 days supply left in the silos. The world teetered on the brink of catastrophe. What would have happened if the crops of the next season had failed? Would those that had food have kept it, leaving those that had not with the choice of starvation or atomic revolution? It did not happen, but the fact that it could have happened was shocking enough to remove any doubt concerning the ideas of Thomas Malthus, so that the majority of the thinking world understood

the global meaning of the Sahel famines of 1973/74. In this region of savannah, which lies in the path of the advancing Sahara, 12 million people had to exist entirely on famine supplies sent in from outside. Many died and the fires of revolution burned once more across the television screens, radios and newspapers of the world.

Other much-publicized happenings lent weight to the cause of Malthus. Many of the advances on which the new world agriculture depended were seen to be two-edged. Rachel Carson wrote *Silent Spring* in which she urged agriculturalists to take a critical look at the proliferation of their own form of chemical warfare, synthetic pesticides. She warned of destruction of wildlife on a massive scale, the creation of imbalance in natural systems, and the threat to the health of man himself.

Concern over atmospheric pollution became action when smog killed 4000 Londoners in 1952. The 'death' of Lake Erie by eutrophication (enrichment) and pollution did the same for the cause to halt water pollution. The Torrey Canyon disaster was just one in a long line of oil slicks which lubricated the Food and Agricultural Organization to convene an international conference in Rome to consider whether man could pollute the other five-sevenths of the earth. Massive evidence began to accrue which showed that not only was man stressing the agricultural potential of the space-ship earth, but he was also beginning to affect the fitness of the life support system on which he depended.

The period saw the birth of many conservation movements who were concerned at the disruption of the natural ecosystems and the extinction of their component plants and animals. Suddenly, people within the affluent societies were beginning to question the cost of their affluence. It was not only those on the right end of the overflowing GNPs, it was a grass-roots movement, the call for conservation coming stridently from all corners of society.

Let us come back to the question of what GNP really measures. In the words of two of the greatest of the new breed of conservationists Paul R. and Anne H. Ehrlich, 'It is not a measure of the degree of freedom of the people of a nation. It is not a measure of the health of a population. It is not a measure of the state of depletion of natural resources. It is not a measure of the stability of the environmental systems upon which life depends. It is not a measure of security from the threat of war. It is not, in sum, a comprehensive measure of the quality of life.'

How do you start to measure the quality of life since no two people's idea of quality will be the same? What we need are hard facts and none could be harder than the facts about the modern methods of agriculture on which much of the hope of past years has depended. Think back to our hunting origins. The hunter used about 0·1 joule of his own energy to catch and gather every joule of his food, and he kept fit while he was doing it. Simple agriculture based on human muscle and staple crops, even today, yields more energy than is required to produce it. However, from this point up the systems require a greater amount of energy to be invested than is produced. It does not matter if the ploughing is done by bullock, horse or tractor–all of them need 'feeding'. Only a few types of farming are shown on the diagram, but from your own knowledge of giant tractors roaring up and down, aeroplanes with spray booms, gigantic silos and combine harvesters, the picture of energy use becomes very clear both in terms of decibels and litres. All the most modern, most productive agricultural systems are absolutely dependent on massive inputs of energy. The portion of the diagram which shows the advance of American agriculture over the 60 years prior to Apollo 11 is very revealing, especially in the light of the fact that the whole world has come to rely more and more on its excess. No wonder Kuwait with massive oil reserves and only about 800000 mouths to feed was second in the 1969 GNPs.

One of the costs of agriculture which is hidden in the straight statistics is the cost of crop and livestock breeding programmes. With the

Below A Western Grebe (*Aechmophorus occidentalis*) killed by an oil spill near San Francisco. This is just one of many millions of sea birds that die this ghastly death each year.

Above A boom in Swedish farming, crop spraying to combat pests and diseases.

The rising cost of farming = the rising cost of living

Left It seems sensible to put as little energy as possible into the production of food which is, after all, only energy in another form. The agricultural methods at the bottom of the scale do just that, but it is very hard work. As you go up the scale the farming becomes less efficient in terms of energy but the work is easier, both for the farmer and the farm animals. We may decry the conditions under which animals are kept in factory farms but the farmers do not grumble and the veterinary physiologists and the growth rates of the animals indicate that the animals are themselves not too upset by their easy standstill life style. The bottom of the scale is alright as long as the system has got cheap willing labour and the top of the scale is alright so long as there is cheap fossil fuel. The compromise is the stuff of evolution.

The rising cost (in energy) of the USA over the past 60 years is shown to highlight the problem.

US food system
1970
1960
1950
1940
1930
1920
1910

joule input to obtain 1 joule output

20 distant fishing
feedlot beef
10
fish protein concentrate
5
intensive eggs
2
coastal fishing
1
dairy farming
low-intensity eggs
0·5
hunting
gathering
0·2
intensive potatoes
low intensity corn
soya beans
0·1
low-intensity potatoes
0·05
wet-rice culture
shifting agriculture
0·02

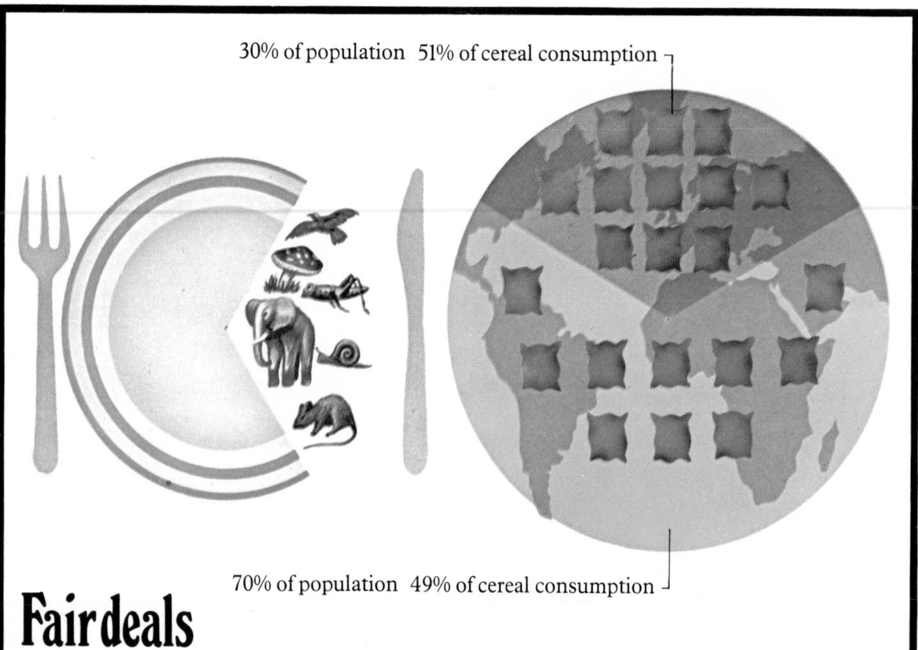

Fair deals

Above left The world of man shares its harvests with a cross-section of the living world. In India, an estimated 30 per cent of the harvest goes to non-human feeders. If we could only stop such post-harvest losses we would solve the immediate problem of feeding all men adequately. Please note that I am referring to post-harvest losses, not the animals' fair share of natural productivity.
Above right The harvests of the world are at present shared out between mankind in a very unfair way. The lady who used to run my favourite fish and chip shop in Kirkby Stephen, northern England, once put it in a nutshell: 'It seems a strange thing that two-thirds of the world are starving while the other third is slimming'. I mumbled my agreement through a mouthful of chips.

Right Super cereals bred to be super productive by the skill of plant breeders and geneticists. This high-yielding variety of wheat is growing in a field in south Bihar, India.

scaling down of the use of pesticides, more attention has been turned to the creation of varieties that are more resistant to both disease and attack by pests. It is a never-ending job in which the plant breeders attempt to keep a large enough jump ahead of the rapid evolution of the pests and parasites.

Massive effort has also gone into the creation of more productive strains of everything. The biggest breakthrough was in the field of 'miracle cereals' and Dr Norman Borlaug was awarded the 1970 Nobel Peace Prize for spearheading what became known as the green revolution. It would be wrong to denigrate the importance of this work in any way but one of the basic rules on which this universe runs states that matter/energy can neither be created nor destroyed, it can only be changed (evolved?) from one form to another. A greater yield must require a greater input of raw materials, which in modern farming means fertilizer and farm machinery, both of which are energy dependent.

It was not Borlaug's fault, or that of any of the dedicated band of thinkers and workers who were striving to make the world a more productive place, that just as the green revolution began to solve age old problems of want, the cost of energy started to climb. This put the price of the new agriculture way beyond the reach of even the super-peasant and the price of world food on a spiral of inflation.

The need for more and more energy spurred on exploitation, exploration and experimentation. Although new technologies, new techniques and massive new reserves are being found, the resources of fossil fuels are finite; they can not go on forever. For this reason a number of the oil-producing states are questioning the use of such a valuable resource of organic chemicals simply as a fuel.

The hopes born at Hiroshima were for a new world powered by energy from the atom–unlimited, unbounded affluence. The cost of development of safe reactors, of

handling the toxic waste, and of ensuring the long-term safety of the core of each plant once it has passed beyond its useful life span, has slowly but surely eroded the promise of cheap power. The question of overall safety of the operation and the possibility of nuclear weapons falling into the hands of unscrupulous terrorist groups fighting for what they consider a just cause has certainly cast grave doubts on the future of atomic power. The dreams are still of unlimited safe power. It may come, but the immediate reality is of a world GNP based on the dwindling resources of oil, gas and coal.

In the last few paragraphs I have rushed over a fantastic range of important happenings and developments. This is done not to hide the facts or bend the truth, but because those facts are so much a part and parcel of your own everyday experience that they should already be a part of your consciousness.

Wherever there is potential it must be used. Twentieth-century man is obeying pre-eminently this rule, which has 'directed' evolution from its physico-chemical beginnings. We use more geochemicals and more water and degrade more energy and more environments than any other species. The potential was there; the rules had to be obeyed; evolution had to happen from photosynthesis through the 'explosion' in the Cambrian to the industrial/medical revolution. Each stage has acted as a positive feedback speeding the process on. The only development that appears to be out of line with the rest of evolution is the new consciousness of concern. Is this the meaning of man's free will?

To date the factors which have limited the process have been external–short-term shortages of energy or raw materials. Now there is an inbuilt control system. The voice of conservation, if heeded, can gain for man, and the way of life which is implicit in the word humanity, the final stamp of evolutionary approval–survival.

'Survival of the fittest' is a dogma which gives us *carte blanche* for the modern rat-race philosophy of life–I am the greatest and I do not care who I have to tread on to get on. But true evolutionary fitness relates to the fitness of playing a role within the society of nature. What is the use of being a peer if there is no one to do your bidding, a student if there is no one with whom to share your knowledge, a pop star if there are no fans who can buy your records, a well-paid miner if there is no industry left to require the coal, or a shop steward in a factory producing the world's top selling car if the midwife who saved your breech-born baby and your wife could not afford the car that got her to the confinement on time. There is no top dog in any living system; each component is part of the whole which is itself much more than the sum of its parts.

The old philosophy of evolution, 'wherever there is potential it will be used' appears equally outmoded by the new consciousness. Now that there is the possibility of restraint we can decide that the potential will only be put to use when it is of long-term benefit to the whole, and that includes all the products of evolution.

Well, perhaps not quite all. Consider, for example, the two notices which can be seen by anyone entering the USA through one of its large airports, one of which states that a smallpox vaccination is required only for those coming from Eritrea and Sudan, while the other warns that the Conservation of Rare Mammals Act forbids the import of certain types of souvenirs. The fight against disease continues to improve the quality of life on a world scale. Perhaps by the time this book is published the smallpox virus will be extinct and international agreement will have stopped the slaughter of all wild creatures that are not directly harmful to man.

One problem is time–can the meek inherit the earth before the rat-race mob ruin it still further? I believe that they can, for at last the meek are becoming vociferous and they are being heard. Let us look at some of the things that are happening on the positive side. The population of a number of countries is beginning to stabilize and even to fall, and it is of interest that all of them are in the top of the GNP bracket. The reason, I must hasten to add, is in no case purely altruistic. It stems from a development of affluent consumerism which has been called the Madison Avenue syndrome. People, it seems, would rather have a new washing machine, a second car, wall-to-wall carpeting and a holiday abroad than a third child. It is a direct spin-off from the rising cost of living. It would be foolish to argue that because its heart is in the wrong place, it is not a good thing. The demand for those goods is boosting certain parts of the present world economy, the reduction in population growth is cutting down the flow of food to those countries, and the spin-off in the technology and sociology of planned parenthood can only benefit the whole world.

The rising cost of food in the developed countries has stimulated research into high-yield crops and into the development of meat substitutes like soya bean. The latter is, without doubt, most important, because in food-chain terms it represents one step closer to the sun, and that means a saving on productive land. Both the seed and the know-how of these new crops are being shared with all nations, thus cutting down not only the actual food gap, but also the ethical gap of two-thirds of the world living on the crumbs which may fall from the rich men's tables.

Movements, like the Friends of the Earth, are making the world conscious of the importance of recycling. Traditional wastes from food and agriculture, and even from the biggest industrial concerns, are being turned into traditional profit. More and more countries are becoming concerned about their wildlife and setting up reserves, refuges and national parks.

There is an overall scaling down in certain development schemes, from grandiose dams and prairie peanuts, to simple ideas of one pump one more affluent peasant, schemes which promise diversity and long-term stability.

Conservation–that is, the sensible utilization of natural resources–is happening; the breakthrough is being made. We must, however, be very

Farmers in Malawi learning to use fertilizer as part of a national campaign to increase crop production.

careful to ensure that this new euphoria is not allowed to lay the ghost of Malthus to temporary rest once more. Time is not on our side, but it never has been on the side of the products of evolution. Remember that time can only be measured against change, and anything which has changed has not, as such, survived. The Lamarckian advance of conscious man is not so time demanding and as it is based in part on written communication, it should at least leave a more detailed, if not as permanent, a record than the chance process of fossilization which pockmarked the path of previous evolution.

The way ahead is clear, at least to those millions who share the new consciousness:

1 A world population, planned in relation to available natural resources and, until that time, a more equitable distribution of those resources to all people, no matter what colour, class or creed.

2 The replacement of GNP as a scale for the success of life by something that has real meaning. A step in the right direction would be to follow the suggestions of the economist Edwin G. Dolan as set out in his smashing little book *Tanstaafl* (*T*here *A*in't *N*o *S*uch *T*hing *A*s *A* *F*ree *L*unch). He argues that GNP should be replaced by two types of GNC (Gross National Cost). The first type of GNC would measure that part of the economy which was based on renewable resources and the recycling of wastes (Gross National Pride). The second type of GNC would be that part which is dependent upon the depletion of non-renewable resources and the production of indestructible wastes (Gross National Shame).

3 A declaration of rights for all wildlife, proclaiming the importance of the role they play in the quality of life and the continued welfare of the biosphere. This must be spearheaded by (a) a total ban on slaughter of all rare animals and plants; (b) the immediate acceleration of the designation of a series of international nature reserves paid for by international money which would preserve a full cross-section of all the living systems extant on earth–a world bank of the successful products of past phases of evolution. Just think of the hardness of future currencies set in terms of nations showing in Gross National Pride and Shame, the role they play in the World Bank of Nature.

4 A scaling up of the efficiency of the use of agricultural land and agricultural produce by (a) decreasing our dependence on animals for our essential amino acids by turning to protein-rich legumes such as soya bean; (b) the more efficient use of organic fertilizers of all sorts; (c) a major attack on harvest losses by pests, which in some countries are destroying about 40 per cent of the crop, so tainting it with toxins produced by spoilage organisms that the food becomes a health hazard.

5 Until such time as a new source of safe energy is found, international effort should be turned to make more use of the current account of world energy–the sun, wind, tides and currents.

6 More of the knowledge and technology now being developed for undersea oil exploration to be turned to the benefit of inshore fish farming. For example, areas already surveyed in south-east Asia could produce high grade protein, equivalent to the contemporary total fish catch of the world's fleets. If only this dream could be turned into reality, the dwindling ocean fish stocks, now being overexploited, could be allowed a complete reprieve.

7 More of the revenue from oil, which is at the moment seeking sound investment, to be used for projects that will turn the tide of desertification–a real investment in the future of botanic man.

I believe that it can be accomplished through reasoned discussion, but I am an optimist speaking from a position of comfortable affluence. Perhaps revolution would appear more appropriate from another viewpoint.

Burchell's Zebra (*Equus burchelli*) was fitted by evolution for life on the dry savannahs of Africa. It is just one of the many products of evolution that has a right to live alongside man.

The Families of Man

Top of the Plots

Gramineae	Rice, wheat, maize, barley, oats, sugar cane, rye, sorghum, millet, all forage grasses
Euphorbiaceae	Cassava (tapioca), chaya
Fabaceae	Beans, soya, peas, lentils, ground nut, tamarugo, alfalfa, clovers, guar
Palmaceae	Coconut, sago, dates, oil palm, pejibe, barbassu, chonta
Araceae	Taro, cocoyam
Dioscoriaceae	Yam
Solonaceae	Potato, tomato, naranjilla
Cucurbitaceae	Squash, cucumber, pumpkin, courgette, melon, buffalo gourd, wax gourds
Apiaceae	Carrot, parsnip, arrechacha, celery, parsley, anise, caraway, rue, dill, fennel, aubergine
Convolvulaceae	Sweet potato
Brassicaceae	Cabbage, cauliflower, turnip, radish, sprouts, rape, broccoli, mustard, kale, kohlrabi, swede, horseradish
Caricaceae	Papaya
Musaceae	Banana, plantain
Moraceae	Bread fruit, mulberry, uvilla, beetroot
Chenopodiaceae	Sugar beet, quinua, sea samphire
Rutaceae	Orange, lemon, grapefruit, tangerine, lime, satsuma, pummelo
Rosaceae	Apple, pear, plum, peach, cherry, almond, apricot, raspberry, loganberry, strawberry, lynganberry
Oleaceae	Olive
Liliaceae	Onion, leek, asparagus, agave
Polygonaceae	Buckwheat, rhubarb
Asteraceae	Lettuce, chicory, endive, artichoke, salsify, scorzonera
Lauraceae	Avocado, bay, cinnamon, sassafras
Bromeliaceae	Pineapple
Ericaceae	Blueberry, cranberry
Caesalpiniaceae	Tamarind
Amaranthaceae	Grain amaranths
Bombacaceae	Durian

Annonaceae	Sour sop, custard apple
Clusiaceae	Mangosteen
Rubiaceae	Coffee
Sterculiaceae	Cocoa
Theaceae	Tea
Vitaceae	Grape
Zingiberaceae	Ginger
Piperaceae	Peppers
Lamiaceae	Mint, thyme, sage
Caprifoliaceae	Elderberry
Orchidaceae	Vanilla
Cannabaceae	Hop
Aceraceae	Maple syrup
Araliaceae	Ginseng, sasaparilla
Juglandaceae	Walnut, pecan
Fagaceae	Sweet chestnut
Corylaceae	Hazelnut
Anarcardiaceae	Cashew nut
Myristicaceae	Nutmeg
Myrtaceae	Cloves
Malvaceae	Okra
Punicaceae	Pomegranate
Irvingiaceae	Wild mango
Urticaceae	Ramie
Potamogetonaceae	Sea grass

Apart from these, which are all flowering plants, the only other plants which are important as a source of food are:

Fungi	Yeast for fermentation (brewing) and baking; mushrooms, truffles and other edible toadstools
Algae	Brown, including giant kelp as a source of alginates; red, nori, carragheen, dulse
Prokaryotes	*Spirulina nostoc*
Ferns	Fiddle heads

Bibliography

Allan, M., *Darwin and His Flowers*, Faber (1977). The author argues very elegantly that it was the plants in Darwin's life that gave him the key to natural selection.

Bellamy, D. J., *Bellamy on Botany*, BBC Publications (1972, revised edition 1975). A primer in very basic botany.

Bellamy, D. J. and Boorer, M., *Green Worlds: Plants and Forest Life* (The Living Earth series), Aldus Books Ltd (1976) Doubleday (1976). Important facts about plants and animals with super pictures.

Bellamy, D. J., *The Life-giving Sea*, Hamish Hamilton (1975), Crown (1975). Covers the world of the sea in breadth if not depth.

Boughey, A. S., *Man and Environment* Collier Macmillan (2nd revised edition, 1975). An excellent textbook covering a lot of important fields.

Brunsden, D. and Doornkamp, J. C. (Ed.), *The Unquiet Landscape*, David and Charles (1974). This is the stuff landscapes are made of.

Carson, R. *Silent Spring*, Hamish Hamilton (1963). The classic work that set the world thinking.

Darwin, C., *Voyage of the 'Beagle'*, Bantam Books (1973). A reprint of the classic that started it all.

Development Forum (editions in English, French, German, Italian and Spanish); published monthly by CESI/OPI United Nations, Palais des Nations, Geneva 10. The facts about botanic man brought up to date and put into perspective.

Dolan, E. G., *Tanstaafl*, Holt, Rinehart and Winston (1969).

Ehrlich, P. R. and A. H., *Population Resources and Environment*, W. H. Freeman (2nd revised edition 1972).

Hardin, G. (Ed.) *Population Evolution and Birth Control*, W. H. Freeman (2nd revised edition 1969). A collage of the most important ideas and action.

Henderson, L. J., *Fitness of the Environment*, Macmillan Co. (1913, reprint 1958). A classic before its time. It needs updating but it also needs reading.

Holden, E., *The Country Diary of an Edwardian Lady*, Michael Joseph Webb and Bower (1977). An entrancing book to own.

Houwink, R., *Sizing Up Science*, John Murray (1975). Hard facts presented in an interesting way.

Jantsch, E., Waddington, C. H. (Ed.), *Evolution and Consciousness*, Addison Wesley (1976). An advanced text, of great importance.

Land, F., *The Language of Mathematics*, John Murray (1960, new edition 1975). A must for anyone who, like me, used to be frightened of mathematics.

Measures, D., *Bright Wings of Summer*, Cassell (1977). For anyone who likes magic and wants to know what it feels like to be a biologist.

Meggers, B. J., *Amazonia: Man and Culture in a Counterfeit Paradise*, (Worlds of Man series). Aldine Publishing Company (1971). One of my favourite books.

Mitchell, J. (Ed.), *The Natural World*. The Mitchell Beazley Joy of Knowledge Library Volume 1. Mitchell Beazley (1977). All the facts about the natural world beautifully displayed. A delightful work of reference.

Odum, H. T. and Odum, E. C., *Energy Basis for Man and Nature*, McGraw Hill (1976). Sizing up the future.

'Scientific American', Readings from: *The Biosphere*, W. H. Freeman (1970); *Oceanography*, W. H. Freeman (1971); *Plant Agriculture*, W. H. Freeman (1970). 'Scientific American' keeps the world abreast of the meteoric advance of science. These are all key works in these key fields.

Tarling, D. H. and M. P., *Continental Drift*, G. Bell 1971 (hardback) Penguin Books 1972 (paperback). Once you have read it you will be able to feel the continents move.

Walter, H., *Vegetation of the Earth in Relation to Climate and the Ecophysiological Conditions.* Heidelberg Science Library Volume 15 Springer-Verlag, New York (1973), English University Press (1974). A little book about a big subject. It reads like a novel.

Weisz, P. and Fuller M., *The Science of Botany*, McGraw Hill (1962). An excellent basic text, giving the facts on botany.

Acknowledgements

A television series should be a work of art, but only the viewing public can make the verdict. The team who shared with me the excitement of the broad canvas of the vegetation of the world and contained its breadth, beauty and importance in 265 minutes of film were:

Sally Barnsley	PRODUCTION ASSISTANT
Randal Beattie	PRODUCER/DIRECTOR (KIWI)
Ted Berry	ELECTRICIAN
Mary Crewe	PRODUCTION ASSISTANT (AND MUM)
Martyn Day	RESEARCHER AND EXPEDITER
Terence Dixon	EXECUTIVE PRODUCER
Graeme Duckham	PRODUCER/DIRECTOR (COBBER)
Michael Feldman	ASSOCIATE PRODUCER (EDUCATION)
Robert Fleming	DIRECTOR
Trefor Hunter	SOUND ASSISTANT
Thomas Hurley	ELECTRICIAN
Martin Lucas	RESEARCHER
Brian Mongini	FILM EDITOR
Peter Patman	ORGANISER
Basil Rootes	SOUND
Alison Ryan	PRODUCTION ASSISTANT
Nick Sargent	ASSISTANT CAMERA
Raymond Sieman	CAMERA
Colin Thompson	FILM EDITOR
John Woods	RESEARCHER

My thanks are due to them for all their hard work, tolerance and companionship. I can only hope that my random ideas were the right media for their skills.

David Bellamy

Photographs

J. Allan Cash Ltd, London 158–159; Ardea Photographics, London 18, 107, 113, 124–125, 164–165, 180, 198–199; Ardea Photographics–Ake Lindau 193; Ardea Photographics–Richard Waller 154–155, 168–169; Aspect Picture Library, London 34–35, 184–185; Biofotos, Farnham–Heather Angel 16–17, 20–21, 68–69, 80–81, 83, 87, 178, 181, 182–183; Biofotos–S. Summerhays back end paper; Biophoto Associates, Leeds 9 left, 37 right, 45, 49, 57, 66, 67; Botanical Enterprises (Publications) Ltd, Bishop Auckland 11, 13, 36–37, 54–55, 74, 82, 84–85, 99, 100, 108–109, 116, 118, 119 bottom, 131, 138 top, 138 bottom, 162 bottom left, 162–163, 166–167, 178–179, 179 top right, 179 bottom right, front jacket flap; Paul Brierley, Harlow 186 top; Centre for Overseas Pest Research, London 48; Bruce Coleman Ltd, Uxbridge–Jen and Des Bartlett 78–79; Bruce Coleman Ltd–Robert Burton 127; Bruce Coleman Ltd–Nicolas Devor 73; Bruce Coleman Ltd–M. P. L. Fogden 15, 21 bottom; Bruce Coleman Ltd–J. Houzel 188; Bruce Coleman Ltd–Norman Myers 192; Bruce Coleman Ltd–Charlie Ott 119 top, 140–141; Bruce Coleman Ltd–Janoslav Poncer 157; Bruce Coleman Ltd–Norman Tomalin 12; Cross Sections, Addlestone–Kim Westerskov 64–65; Cross Sections–Bill Wood 174–175, 176–177; Department of Biology, University of York 9 top right, 9 centre right, 9 bottom right, 52, 70, 186 bottom; FAO Photo, Rome 189, 190–191, 194–195, 197; Pat Morris Photographics, Ascot 47, 50–51, 59; Dr M. Muir, Canberra 32; Natural History Photographic Agency, Saltwood 162 top; Oxford Scientific Films Ltd, Oxford 43; Popperfoto, London 26 top, 26 bottom; David Redfern, London 40–41; RIDA Photo Library, Norbiton–R. T. J. Moody 150–151; G. R. Roberts, Nelson 10, 77, 88–89, 101, 114–115; Royal Botanic Gardens, Kew 60–61; Bryan Sage, Potters Bar 94–95, 128–129, 133 top, 133 bottom, 134–135, 136 left, 136 right, 143 top left, 143 top right, 143 bottom left, 143 bottom right; Mireille Vautier, Paris 144–145; Mireille Vautier–Decool 23, 53, front endpapers; Mireille Vautier–De Nanxe 137; Mireille Vautier–Rizière title spread; Z.E.F.A., London–I. Csenyi-Simonis 160–161; Günter Ziesler, Munich 111, 120–121.

Artwork

Page 7–after Heinrich Walter, *Vegetation of the earth in relation to climate and the eco-physiological condition*, Springer Verlag, 1973; p22–after Betty J. Meggers, *Amazonia, man and culture in a counterfeit paradise*, Aldine Publishing Co., 1971; p123–after Odum *Ecology*, Holt Rinehart and Winston; p149 lower–after *Population Bulletin* vol. 18, no. 1, p4, Feb 1962; p172 top–after F. W. Preston, 'The economical distribution of commonness and rarity', *Ecology* 43: 185–215, 410–432, Copyright © 1962 by the Ecological Society of America; p172 middle and bottom, p173–after R. H. MacArthur and E. O. Wilson, *The Theory of Island Biogeography*, Princeton University Press, 1967; p189 top, p193 bottom, p194–after Kate Simunek, 'Facts on Food', *Development Forum*, 1974.

The publishers are also grateful to Betty J. Meggers for permission to use her data in the table on p23.

Index

Numbers in italic refer to illustrations.